Journal of
Neural
Transmission

Supplementum 30

C. G. Gottfries and S. Nakamura (eds.)

Neurotransmitter
and Dementia

Springer-Verlag Wien GmbH

Prof. Dr. C. G. Gottfries
Department of Psychiatry and Neurochemistry, Gothenburg University, Hisings Backa, Sweden

Prof. Dr. S. Nakamura
3rd Division of Internal Medicine, Hiroshima University Hospital, Hiroshima, Japan

Originally published by Springer-Verlag Wien in 1990

Printed on acid-free paper

With 44 Figures

ISBN 978-3-211-82190-9   ISBN 978-3-7091-3345-3 (eBook)
DOI 10.1007/978-3-7091-3345-3

Preface

Since three groups revealed a decrease in choline acetyltransferase activity in the Alzheimer brain in 1976, much attention has been focussed on neurotransmitter abnormality in Alzheimer's disease. Neurotransmitters are important in neuropsychiatric diseases because of their role in the pathogenesis of these diseases; at the same time, drugs that influence their metabolism may have particular significance in the treatment. In the past ten or more years, hundreds of reports have appeared on the abnormalities of various neurotransmitters in Alzheimer's disease although therapeutic trials on Alzheimer's disease through neurotransmitters have not hitherto been successful. Recently, strategies on neurotransmitter abnormality have widened as a result of new methodologies of molecular biology, including elucidation of structures of receptor-channels, receptor-mediated polyphosphoinositide metabolism or metabolism of enzymes related to neurotransmitters.

Many efforts have been made in the search for new openings in neurotransmitter research on Alzheimer's disease. It is timely, therefore, that Prof. Hasegawa, the president of the Fourth Congress of the International Psychiatric Association (IPA) has arranged publication of the Symposium "Neurotransmitter and Dementia" (held at the IV IPA Congress, Tokyo, September 6–8, 1989) as a Supplement to the "Journal of Neural Transmission". We have edited articles presented at the symposium and we are sure that this book will help to lead the way to a therapeutic breakthrough as well as clarification of the disease.

Finally we would like to express our many thanks to Prof. Dr. A. Carlsson, Managing Editor and Prof. Dr. P. Riederer, Coordinating Editor of "Journal of Neural Transmission", and also to Springer-Verlag Wien, especially Mr. Frank Chr. May.

C. G. Gottfries
S. Nakamura

Contents

Goto, T., Kuzuya, F., Endo, H., Tajima, T., Ikari, H.: Some effects of CNS cholinergic neurons on memory 1

Nakamura, S., Kawashima, S., Nakano, S., Tsuji, T., Araki, W.: Subcellular distribution of acetylcholinesterase in Alzheimer's disease: abnormal localization and solubilization 13

Ikeda, Y., Okuyama, S., Fujiki, Y., Tomoda, K., Ohshiro, K., Itoh, T., Yamauchi, T.: Changes of acetylcholine and choline concentrations in cerebrospinal fluids of normal subjects and patients with dementia of Alzheimer-type 25

Gottfries, C. G.: Disturbance of the 5-hydroxytryptamine metabolism in brains from patients with Alzheimer's dementia 33

Lee, S., Chiba, T., Kitahama, T., Kaieda, R., Hagiwara, M., Nagazumi, A., Terashi, A.: CSF β-endorphin, HVA and 5-HIAA of dementia of the Alzheimer type and Binswanger's disease in the elderly 45

Minthon, L., Edvinsson, L., Ekman, R., Gustafson, L.: Neuropeptide levels in Alzheimer's disease and dementia with frontotemporal degeneration . . . 57

Shimohama, S., Ninomiya, H., Saitoh, T., Terry, R. D., Fukunaga, R., Taniguchi, T., Fujiwara, M., Kimura, J., Kameyama, M.: Changes in signal transduction in Alzheimer's disease 69

Subject Index 79

Listed in Current Contents

J Neural Transm (1990) [Suppl] 30: 1–11

Some effects of CNS cholinergic neurons on memory

T. Goto, F. Kuzuya, H. Endo, T. Tajima, and **H. Ikari**

Department of Geriatrics, Nagoya University School of Medicine, Nagoya, Japan

Summary. The aim of this study is to observe the relationship between the impairment in passive avoidance task induced in rats by the i. p. administration of muscarinic antagonists, scopolamine and methyl-scopolamine, and the change in acetylcholine (ACh) output induced by these drugs. Initially we studied the effects of these drugs on the animals' performance of a step-through passive avoidance task. We then measured the change in ACh levels after administration of these drugs using an in vivo brain dialysis technique. Scopolamine was effective in impairing the performance of the passive avoidance task, while methyl-scopolamine did not have clear effects on the performance of the task. With regard to ACh output, scopolamine increased ACh dose-dependently and methyl-scopolamine also affected ACh release. These data suggest that the accumulation of ACh in the synaptic cleft may be involved in the memory deficit induced by scopolamine.

Introduction

Central nervous system (CNS) cholinergic neurons are believed to be involved in learning and memory. In addition, the CNS cholinergic neurons are of additional interest since it has been suggested that Alzheimer's disease, which is accompanied by a profound deficit in memory and cognitive function, is associated with a deficiency of ACh in the cholinergic neurons. It has been demonstrated that in Alzheimer's disease there is a decline in the cortical levels of choline acetyltransferase (ChAT), an enzyme involved in the synthesis of ACh and a specific marker for cholinergic neurons. Cholinergic depletion has been shown to be most severe in hippocampus, parietal and temporal cortex (Davies and Maloney, 1976; Perry et al., 1977). Furthermore, the reduction in cortical ChAT has been shown to correlate with the severity of intellectual decline in life (Perry et al., 1978).

Pharmacological manipulation of central cholinergic systems can produce significant changes in the performance of several types of tasks in different animal species and in humans. For example, scopolamine, a centrally-acting muscarinic blocker, is widely used to impair learning and memory in humans (Drackman and Leavitt, 1974; Drackman, 1977; Peterson, 1979; Caine et al., 1981; Beatty et al., 1986; Broks et al., 1988), marmosets (Ridley et al., 1984) and rodents (Higashida and Ogawa, 1986; Elrod and Buccafusco, 1988; Peele, 1988; Rush, 1988), although its precise mechanism of action is still unknown. Whereas methyl-scopolamine, a peripherally-acting muscarinic antagonist, does not pass through the blood-brain barrier (Blozovski and Hennocq, 1982; Hiraga and Iwasaki, 1984). The present study is the first, to our knowledge, in which the dynamic changes of ACh in freely moving rats administered scopolamine or methyl-scopolamine are correlated with an impairment in completing a passive avoidance task.

Material and methods

Behavioral testing

We constructed a two-compartment shuttle box of the dimensions shown in Fig. 1. Male Wistar rats weighing 220 g were obtained from Keary (Nagoya, Japan) and housed with free access to food and water. A rat was placed in the lighted compartment of the box. As soon as it entered the dark compartment, an electric shock, A.C. 1 mA/1 sec, was delivered to the feet through the grid. In the retention trial following 24 hr of acquisition, 4 groups of rats received either saline, 0.02, 0.2 or 2 mg/kg scopolamine and 2 mg/kg methyl-scopolamine intraperitoneally, respectively. Retention of the passive avoidance task was measured by replacing a rat in the lighted compartment and measuring the time latency to enter the dark compartment before drug administration (preinjection) and 1 min, 40 min and 24 hr after drug administration. The test ended either a) when the rat entered the dark compartment or b) when 300 sec had elapsed.

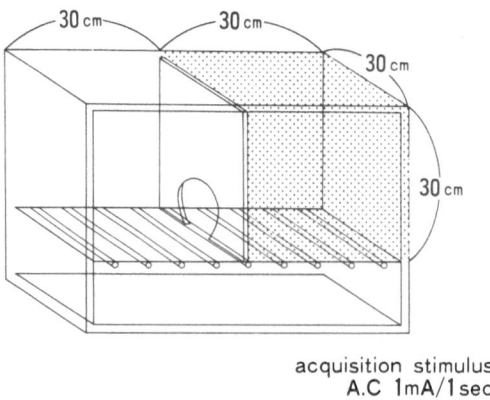

acquisition stimulus
A.C 1mA/1sec

Fig. 1. Two-compartment shuttle box

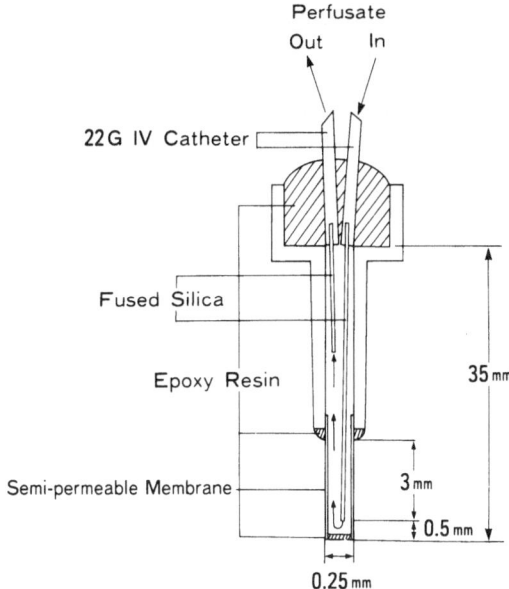

Fig. 2. Diagram of the in vivo brain dialysis probe

Brain dialysis procedure

Male Wistar rats weighing 220 g were anesthetized with sodium pentobarbital (50 mg/kg i.p.; Pitman-Moore, U.S.A.) and a guide cannula was implanted in the left striatum for the subsequent insertion of a dialysis probe. The stereotaxic coordinates for placement in the striatum were 1.6 mm anterior to the bregma, 3.5 mm lateral to the midline, and 3 mm ventral to the dura according to the atlas of Pellegrino and Cushman (1981). Following implantation, the guide cannula was firmly fixed to the skull with two anchor screws and dental cement, and a dummy probe was inserted into the guide cannula. The dummy probe was left in the brain until the dialysis experiment was started. The dialysis probe was constructed with a semipermeable membrane (MW cutoff of 5000, Nikkiso, Tokyo, Japan), fused-silica tubing (0.03 mm inner diameter, Shinwa Kako Co., Ltd., Kyoto, Japan) und 22 gauge intravenous catheter (Critikon, Tokyo, Japan) using the technique of Ozaki et al. (1987) (Fig. 2). We allowed at least two days to pass after the operation, to avoid the effects of anesthesia, then, the dialysis probe was gently inserted into the guide cannula and was subsequently perfused with Ringer's solution (147 mM Na^+, 2.3 mM Ca^{2+}, 4 mM K^+ 155.6 mM Cl^-) containing 100 µM physostigmine, an inhibitor of acetylcholine esterase (AChE), at a flow rate of 2 µl/min (Damsma et al., 1987; Ajima and Kato, 1987). Scopolamine and methyl-scopolamine, dissolved in 1 ml of saline, were administered intraperitoneally 3 hr from the beginning of the dialysis experiment.

Biochemical analysis of perfusate

The concentration of ACh in the perfusate was determined by high-performance liquid chromatography (HPLC)-electrochemical detection (ECD) with the enzyme-column on which acetylcholine esterase and choline oxidase were immobilized (Eicom

Co., Ltd., Kyoto, Japan). The mobile phase was 0.05 M sodium phosphate buffer, pH 8.0, containing 1 g/l sodium 1-octanesulfonic acid (SOS) and 1 mM tetra-methylammonium (TMA). The perfusate was collected for 20-min intervals and each 20 μl of the perfusate to which were added 10 pmol of ethyl-homocholine (EHC), an internal standard, was applied on the HPLC-ECD apparatus.

Materials

Scopolamine hydrochloride, scopolamine methyl bromide, eserine (physostigmine) hemisulfate and SOS were obtained from Sigma (St. Louis, U.S.A.). EHC was a gift from Eicom. All other chemicals used were of analytical grade.

Results

Passive avoidance task

As shown in Fig. 3, scopolamine was effective in impairing the performance of the passive avoidance task measured either as an increase in the percentage of avoidance failure (left panel), or as a decrease in step-through latency (right panel). During the acquisition experiment, the initial latencies for entering the dark compartment did not differ significantly among the four groups of the rats. During the retention trial, no rat entered the dark compartment within 300 sec during preinjection and 24 hr postinjection. Scopolamine administration increased avoidance failure in a dose dependent fashion. This effect was accompanied by a scopolamine-induced decrease in latency to a significant extent as shown in the right panel of Fig. 3.

As shown in Fig. 6, passive avoidance task was not impaired by methyl-scopolamine. The effects of methyl-scopolamine were completely the same as those of saline. This result was in agreement with the previous studies.

Measurement of ACh

A typical elution pattern on HPLC-ECD was shown in Fig. 4. Following the front peak, Ch, EHC and ACh were eluded. Figure 5 shows the effects of several doses of scopolamine on ACh output. As shown in the left upper panel, saline injection did not increase ACh release. Although the concentration of ACh in the perfusate obtained every 20 min differed among animals, levels became stable within 3 hr from the beginning of the dialysis. The average of three previous values before injections were standardized as 100 for each individual animals. ACh release was directly related to the amount of scopolamine injected (Fig. 5). The results illustrated in Fig. 5 are summarized in Table 1. The amount of ACh collected increased to

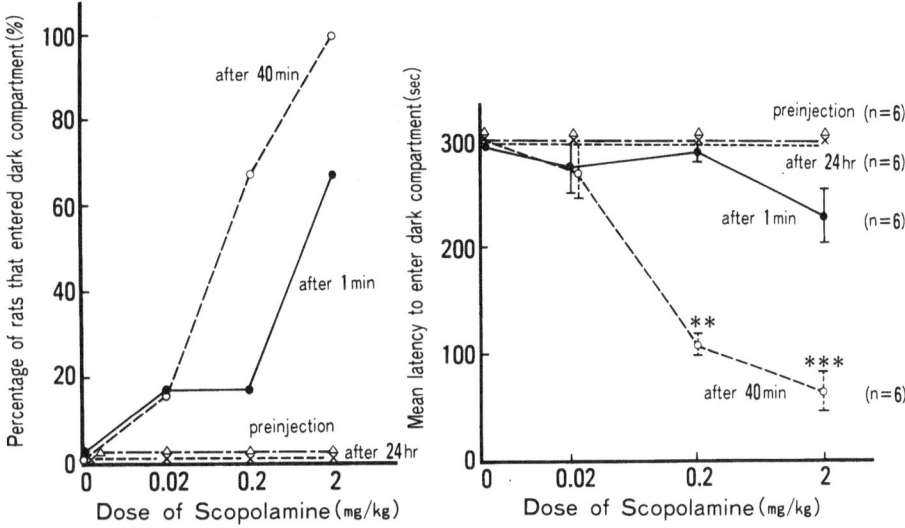

Fig. 3. Effects of scopolamine IP injection on percentage (left) and latency (right) of entering the dark compartment in retention trial. Results in right panel are shown as mean ± S.E.M. Statistical significance of the differences was judged by Student's t-test: ** p < 0.01; *** p < 0.001. Number of samples examined is given in parentheses

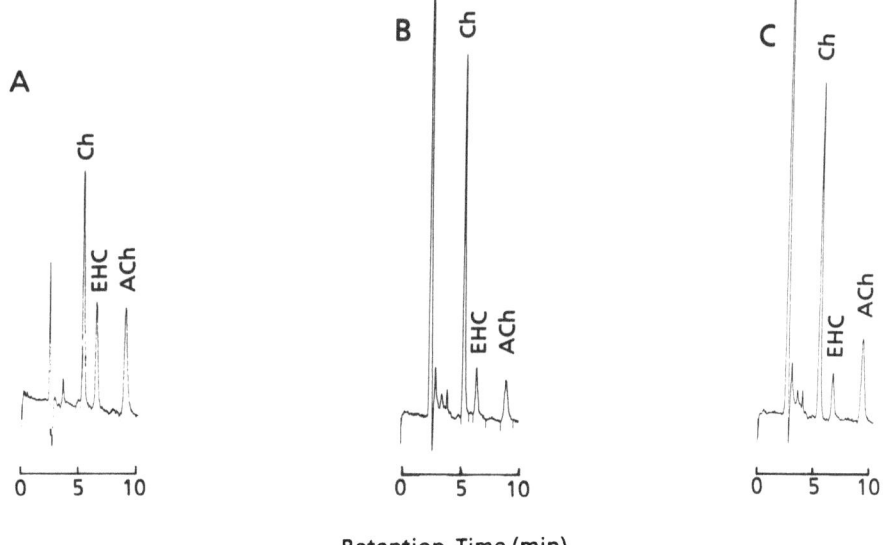

Fig. 4. Chromatograms of acetylcholine related compounds in standard solution or in the dialysate of the rat striatum. **A** Standard solution contained 10 pmol of Ch, ACh and EHC. **B, C** Before and after scopolamine administration

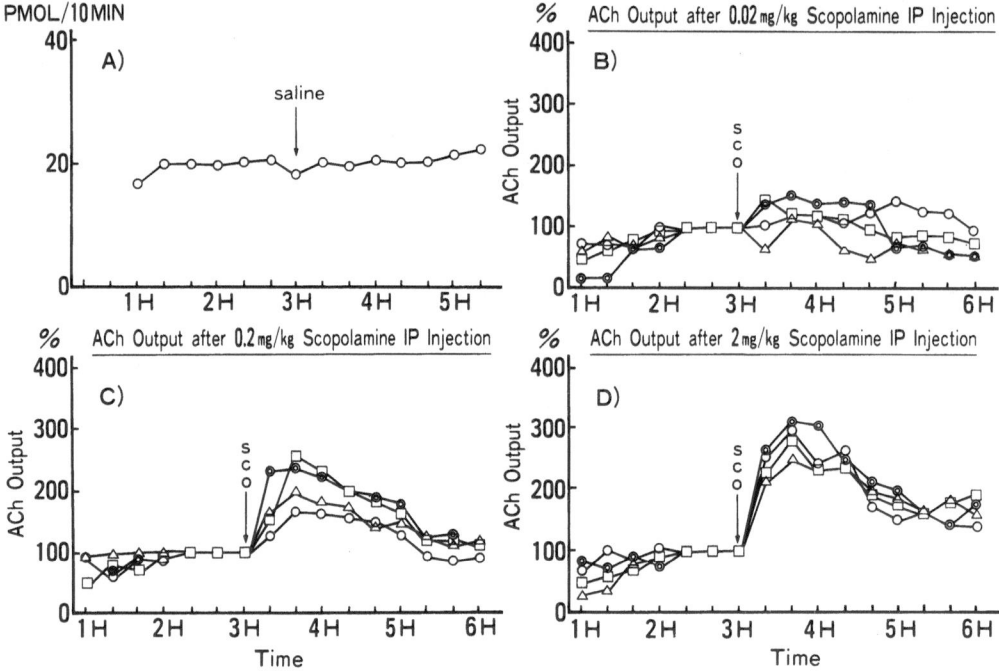

Fig. 5. Effects of scopolamine IP injection on acetylcholine output. **A** shows typical time courses of acetylcholine output in one rate. Saline did not increase acetylcholine output

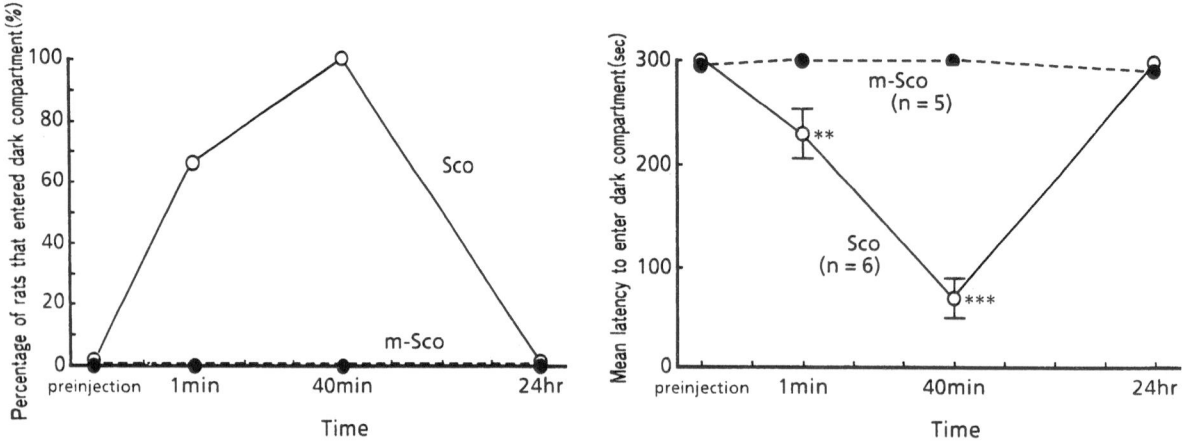

Fig. 6. Comparison of effects of scopolamine and methyl-scopolamine on percentage (left) and latency (right) of entering the dark compartment in rentention trial. Results in right panel are shown as mean ± S.E.M. Statistical significance of the differences was judged by Student's t-test: ** p < 0.01; *** p < 0.001. Number of samples examined is given in parentheses

Table 1. Summary of effects of scopolamine IP injection on acetylcholine output

Dose of scopolamine	1 h			2 h			3 h			4 h			5 h			6 h
0.02 mg/kg (n=4)	50.0 ±10.3	61.7 ±13.1	74.1 ±2.6	86.1 ±5.9	100	100	100	113.4 ±14.7 *	126.0 ±7.6	120.0 ±6.7	107.0 ±14.0	102.4 ±16.6	92.0 ±14.2	86.6 ±12.4	88.9 ±12.0	70.6 ±7.9
0.2 mg/kg (n=4)	73.2 ±10.3	78.9 ±7.0	88.9 ±5.5	87.9 ±2.6	100	100	100	171.5 ±18.8 *	214.7 ±16.6 **	199.8 ±13.6 **	182.6 ±8.4 **	164.5 ±10.4 *	154.8 ±9.5 **	116.6 ±7.5 **	112.0 ±7.6 **	112.4 ±6.7 **
2 mg/kg (n=4)	57.1 ±10.5	68.2 ±11.5	80.7 ±3.4 *	91.5 ±4.6	100	100	100	241.5 ±9.9 ***	285.8 ±11.5 ***	251.8 ±15.5 ***	247.5 ±5.5 ***	194.0 ±7.5 **	178.3 ±8.7 **	163.8 ±1.0 **	162.8 ±8.1 **	168.3 ±8.8 ***

Data are shown as mean±S.E.M. Statistical significance of the differences was judged by Student's t-test: * $p < 0.05$; ** $p < 0.01$; *** $p < 0.001$ (vs 0.02 mg/kg). Number of samples examined is given in parentheses

Table 2. Summary of effects of scopolamine and methyl-scopolamine on acetylcholine output

Dose of scopolamine	1 h			2 h			3 h			4 h			5 h			6 h
Methyl-scopolamine 2 mg/kg (n=3)	61.0 ±10.7	80.2 ±6.1	92.6 ±7.9	100.1 ±4.1	100	100	100	111.7 ±5.7	118.1 ±12.8	128.3 ±10.5	113.2 ±0.6	119.8 ±3.2	102.0 ±2.4	106.0 ±1.7	103.4 ±2.1	102.0 ±0.5
Scopolamine 2 mg/kg (n=4)	57.1 ±10.5	68.2 ±11.5	80.7 ±3.4	91.5 ±4.6	100	100	100	241.5 ±9.9 ***	285.8 ±11.5 ***	251.8 ±15.5 ***	247.5 ±5.5 ***	194.0 ±7.5 ***	178.3 ±8.7 ***	163.8 ±1.0 ***	162.8 ±8.1 **	168.3 ±8.8 **

Data are shown as mean±S.E.M. Statistical significance of the differences was judged by Student's t-test: ** $p < 0.01$; *** $p < 0.001$ (vs methyl-scopolamine). Number of samples examined is given in parentheses

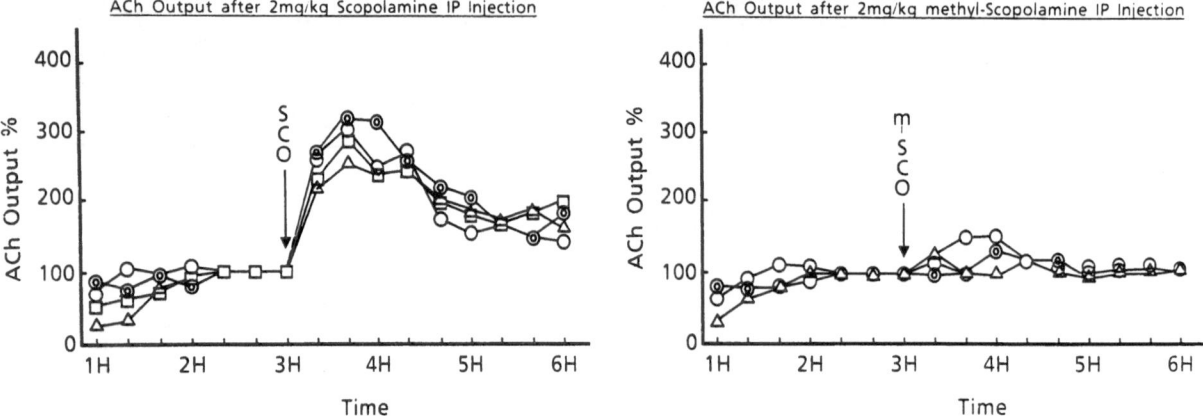

Fig. 7. Comparison of effects of scopolamine and methyl-scopolamine on acetylcholine output

285% of the control level at 2 mg/kg, and 215% at 0.2 mg/kg, and 125% at 0.02 mg/kg. Dose of 2 mg/kg or 0.2 mg/kg scopolamine significantly increased ACh output.

As shown in Fig. 7, the ACh levels were slightly affected by 2 mg/kg methyl-scopolamine. The results shown in Fig. 7 are summarized in Table 2. The effect of 2 mg/kg methyl-scopolamine on ACh release was similar to that of 0.02 mg/kg scopolamine.

Discussion

It is generally accepted that cholinergic blockers stimulate the release of ACh; however, the precise mechanism of action is still unclear. It has been suggested that intracerebral muscarinic ACh receptors are divided into two subtypes, postsynaptic M_1 receptors and presynaptic M_2 receptors, which are located in nerve terminals and regulate the release of ACh (Marchi et al., 1983; Mash et al., 1985). Suzuki (1988) suggested the presence of an M_1 muscarinic autoreceptor that modulates the spontaneous release of ACh in the rat basal forebrain. One possible mechanism for ACh release by receptor blockers is that they act not only on the postsynaptic receptors but also on the presynaptic autoreceptors to inhibit the feedback inhibition of ACh release by those autoreceptors.

Although cerebral cortex or hippocampus is thought to be more concerned with learning and memory, we measured ACh content in striatum since ACh content in former areas is about half (Potter, 1983) or one-third (our own unpublished experiments) of that in striatum and ACh concentration is sometimes undetectable in the cerebral cortex or hippocampus.

Two drugs, physostigmine and scopolamine, each of which increased the concentration of ACh in the perfusate, were used in the present experiments. It was impossible to detect ACh in the perfusate without physostigmine in the perfusing fluid because the sensitivity for ACh of our HPLC apparatus was 2 pmol. The effect of scopolamine was studied after 3 hr from the beginning of the dialysis experiment, when the effect of physostigmine had become stable. The present experiment demonstrated that the effects of scopolamine on a passive avoidance task and ACh release were dose-dependent.

Treatment with 2 mg/kg scopolamine caused a 100% impairment in the passive avoidance task. These results agree with those reported previously (Hiraga and Iwasaki, 1984; Higashida and Ogawa, 1987; Elrod et al., 1988; Peele, 1988). However, the rats administered 2 mg/kg scopolamine became restless in our experiment. Similarly Ridley et al. (1984) reported that a dose of 0.1 mg/kg scopolamine induced agitation in marmosets. Damsma et al. (1987) also reported that atropinized (10 µmol/kg) rats were restless, whereas control rats usually returned to their sleeping posture within 10 min after they were injected. Rush (1988) also observed that a dose of 3 mg/kg scopolamine induced an increased frequency of dark compartment entries and thought that other compounds which increased motor activity, e.g., amphetamine, might also induce deficits in passive avoidance if this increased frequency is resulted from an induced hyperactivity. He concluded that future studies might well focus on this issue because amphetamine had little, if any, impairing effect on retention test performance. Weiner (1985) pointed that scopolamine in therapeutic doses normally caused drowsiness, euphoria, amnesia, fatigue, and dreamless sleep, whereas excitement, restlessness, hallucinations, or delirium occurred regulary after large doses of scopolamine.

It is therefore difficult to conclude the main reason for rats entering the dark compartment following high-dose scopolamine administration, whether it is a true memory disturbance or confusion due to an over-release of ACh. It is known that the memory disturbance in senile dementia of Alzheimer type (SDAT) patients is partly due to a deficiency of cholinergic neurons. In contrast, scopolamine-induced amnesia may result from an excess of ACh. It is equivocal to use scopolamine amnesia as a model of SDAT without considering the dose. In addition, from the clinical viewpoint, the memory disturbance induced by scopolamine differs from that of SDAT (Beatty et al., 1986).

Furthermore, the results of the present experiments demonstrate that large doses of methyl-scopolamine administered may pass through the blood-brain barrier and influence ACh release. Weiner (1985) has pointed out that since quaternary compounds penetrate the blood-brain barrier poorly, antimuscarinic drugs of this type show little in the way of central effects. The results of the present study propose a possibility that 1%

of methyl-scopolamine can penetrate the blood-brain barrier and is effective in ACh release since the pattern and magnitude of ACh release induced with 2 mg/kg methyl-scopolamine is similar to those induced with 0.02 mg/kg scopolamine.

It is concluded that the dose of scopolamine appears to play an important role in the amnesia induced by this drug. The development of more selective blockers of either postsynaptic M_1 receptors or of presynaptic M_2 receptors may help us to understand the role of ACh and scopolamine in amnesia and dementia.

References

Ajima A, Kato T (1987) Brain dialysis: detection of acetylcholine in the striatum of unrestrained and unanesthetized rats. Neurosci Lett 81:129–132

Beatty WW, Butters N, Janowsky DS (1986) Patterns of memory failure after scopolamine treatment: implications for cholinergic hypothesis of dementia. Behav Neural Biol 45:196–211

Blozovski D, Hennocq N (1982) Effects of antimuscarinic cholinergic drugs injected systematically or into the hippocampo-entorhinal area upon passive avoidance learning in young rats. Psychopharmacology 76:351–358

Broks P, Preston GC, Traub M, Poppleton P, Ward C, Stahl SM (1988) Modelling dementia: effects of scopolamine on memory and attention. Neuropsychologia 26:686–700

Caine E, Weingartner D, Ludlow L, Cudahy EA, Wehry S (1981) Qualitative analysis of scopolamine-induced amnesia. Psychopharmacology 74:74–80

Damsma G, Westerink BHC, de Vries JB, Van den Berg CJ, Horn AS (1987) Measurement of acetylcholine release in freely moving rats by means of automated intracerebral dialysis. J Neurochem 48:1523–1528

Davies P, Maloney AJ (1976) Selective loss of central cholinergic neurons in Alzheimer's disease. Lancet ii:1403

Drachman DA, Leavitt J (1974) Human memory and the cholinergic system. Arch Neurol 30:113–121

Drachman DA (1977) Memory and cognitive function in man: does the cholinergic system have a specific role? Neurology 27:783–790

Elrod K, Buccafusco JJ (1988) An evaluation of the mechanism of scopolamine-induced impairment in two passive avoidance protocols. Pharmacol Biochem Behav 29:15–21

Higashida A, Ogawa N (1987) Differences in the acquisition process and the effect of scopolamine on radial maze performance in three strains of rats. Pharmacol Biochem Behav 27:483–489

Hiraga Y, Iwasaki T (1984) Effects of cholinergic and monoaminergic antagonists and tranquilizers upon spatial memory in rats. Pharmacol Biochem Behav 20:205–207

Marchi M, Paudice P, Caviglia A, Raiteri M (1983) Is acetylcholine release from striatal nerve ending regulated by muscarinic autoreceptors? Eur J Pharmacol 91:63–68

Mash DC, Flynn DD, Potter LT (1985) Loss of M_2 muscarine receptors in the cerebral cortex in Alzheimer's disease and experimental cholinergic denervation. Science 228:1115–1117

Ozaki N, Nakahara D, Kaneda N, Kiuchi K, Okada T, Kasahara Y, Nagatsu T (1987) Acute effects of 1-methyl-4-phenylpyridium ion (MPP$^+$) on dopamine and serotonin metabolism in rat striatum as assayed in vivo by a micro-dialysis technique. J Neural Transm 70:241–250

Peele DB (1988) Effects of selection delays on radial maze performance: acquisition and effects of scopolamine. Pharmacol Biochem Behav 29:143–150

Pellegrino LJ, Pellegrino AS, Cushman AJ (1981) A stereotaxic atlas of the rat brain, 2nd edn. Plenum Press, New York London

Perry EK, Perry RH, Blessed G, Tomlinson BE (1977) Necropsy evidence of central cholinergic deficits in senile dementia. Lancet i:189

Perry EK, Tomlinson BE, Blessed G, Bergmann K, Gibson PH, Perry RH (1978) Correlation of cholinergic abnormalities with senile plaques and mental test scores in senile dementia. Br Med J 2:1457–1459

Potter PE, Meek JL, Neff NH (1983) Acetylcholine and choline in neuronal tissue measured by HPLC with electrochemical detection. J Neurochem 41:188–194

Ridley RM, Bowes PM, Baker HF, Crow TJ (1984) An involvement of acetylcholine in object discrimination learning and memory in the marmoset. Neuropsychologia 22:253–263

Rush DK (1988) Scopolamine amnesia of passive avoidance: a deficit of information acquisition. Behav Neural Biol 50:255–274

Suzuki T, Fujimoto K, Oohata H, Kawashima K (1988) Presynaptic M$_1$ muscarinic receptor modulates spontaneous release of acetylcholine from rat basal forebrain slices. Neurosci Lett 84:209–212

Weiner N (1985) Atropine, scopolamine, and related antimuscarinic drugs. In: Goodman LS, Gilman AG (eds) The pharmacological basis of therapeutics, 7th edn. Macmillan, New York, pp 130–144

Authors' address: Dr. T. Goto, Department of Geriatrics, Nagoya University School of Medicine, 65 Tsuruma-Cho, Showa-Ku, Nagoya, 466 Japan.

J Neural Transm (1990) [Suppl] 30: 13–23

Subcellular distribution of acetylcholinesterase in Alzheimer's disease: abnormal localization and solubilization

S. Nakamura, S. Kawashima, S. Nakano, T. Tsuji, and **W. Araki**

Department of Neurology, Faculty of Medicine, Kyoto University, Kyoto, Japan

Summary. AChE activity was detected mainly in membrane-bound fractions in the frontal cortex of autopsied control or Alzheimer brain as well as rat cerebral cortex. However, the distribution of AChE among various membrane fractions was different between control and Alzheimer brains. The highest specific activity was detected in the fraction enriched with senile plaque, which was obtained from the Alzheimer brain by sonication, solubilization with detergent and centrifugation on a sucrose density gradient. The senile plaque enriched fraction was incubated with purified collagenase or protease and centrifuged at 100,000 g for 60 min. More than 50% of AChE activity was detected in the supernatant fraction. AChE in the supernatant solution showed a property of G4 isozyme. AChE might probably be anchored to the senile plaque through its collagen tail and be solubilized with collagenase or protease, producing a G4 isozyme.

Introduction *

Alzheimer's disease affects the cholinergic system of the brain depleting the cholinergic marker enzymes, including choline acetyltransferase and acetylcholinesterase (AChE) (Davies and Maloney, 1976; Perry et al., 1977; Perry, 1980; Koshimura et al., 1986; Younkin et al., 1986; Hammond and Brimijoin, 1988). Moreover, histochemical studies on AChE distribution in the cerebral cortex have revealed a difference between control subjects and Alzheimer patients (Friede, 1965; Mesulam and Moran, 1987). However, the abnormal distribution and properties of AChE have not been examined in detail by a quantitative analysis. The neurochemical research is also required to determine an accurate rate of AChE to butyrylcholinesterase (BuChE) activity in Alzheimer brain, and to elucidate the

---

* Abbreviations used: *AChE* acetylcholinesterase, *BuChE* butyrylcholinesterase

nature of AChE, a mechanism of the difference in subcellular distribution or an effect of AChE inhibitor (Mesulam et al., 1987; Geula and Mesulam, 1988). The present paper describes the subcellular distribution of AChE in rat brain, autopsied control or Alzheimer brain together with the solubilization from the fraction enriched with senile plaque.

Materials and methods

Subjects selection

Autopsied human brains were obtained from 6 patients with non-neurological diseases (3 females and 3 males, mean age 83.7 ± 2.7 SEM) and 8 patients with Alzheimer's disease (5 females and 3 males, mean age 82.3 ± 3.4 SEM). There was no significant difference in autopsy delay between the 2 groups: the mean delays for the control group and Alzheimer group were 7.8 and 7.1 h, respectively. Brains were cut into 2 hemispheres at autopsy. One hemisphere was fixed in 10% formalin for pathological examination. The other was immediately frozen with liquid nitrogen and stored at $-80\,°C$ before use. The hemisphere stored at $-80\,°C$ was kept at $-20\,°C$ for 16 h and then at $0\,°C$ for 2–3 h. Then we removed the arachnoid membrane and pial vessels with perforating arteries, and cut out a block from the frontal cortex (Brodmann area 10) for the study on AChE.

The diagnosis of patients was made by clinical and radiological findings, and pathological examinations. All patients examined as Alzheimer's diseases were clinically moderate or severe at FAST stage 5–7 (Reisberg, 1986) and had no episode of stroke. These cases showed a moderate to severe cortical atrophy on CT scan or MRI without a focal lesion. Morphological examinations disclosed a remarkable appearance of both senile plaques and Alzheimer neurofibrillary tangles in all brains diagnosed as Alzheimer's disease. Mixed-type dementia and equivocal cases were omitted in the present study.

Animals

Male Wistar rats weighing 200 g (4 months old) were sacrificed by decapitation. The brain was immediately removed and then the cerebral cortex was separated from other parts. We prepared subcellular fractions either from the fresh cerebral cortex or from the cortex which had been frozen at $-80\,°C$ for 1 month and thawed before the fractionation.

Preparation of subcellular fractions

The cerebral cortex (0.5 g) was minced at $4\,°C$ and gently homogenized in 5 ml of 0.32 M sucrose containing 10 mM Tris-acetate (pH 7.3) using a loosely fitted Teflon homogenizer. Nuclei and cell debris (P1), crude mitochondria (P2), microsomes (P3), soluble fraction (S3), myelin fragments (A), nerve ending particles (B), and mitochondrial fraction (C) were prepared by the method of Gray and Whittaker (1962). The frontal cortex of autopsied brain (1.0 g) was homogenized and the subcellular fractionation was performed by the same procedure described above.

The fraction enriched with senile plaque was prepared from the frontal cortex of autopsied Alzheimer brains according to the method described by Candy et al. (1986) with modifications. The block (5 g) cut out from the frontal cortex of Alzheimer brain was homogenized in 25 ml of 0.1 M potassium phosphate buffer (pH 7.0) for 30 sec using Polytron homogenizer. The homogenate was sonicated for 1 min with a Raytheon sonic oscillator and centrifuged at 20,000 g for 15 min. The supernatant solution (S'1) was separated from the precipitate (P'1). The precipitate (P'1) was suspended in 15 ml of 2% sodium laurylsulfate and homogenized again for 30 sec using Polytron homogenizer. We centrifuged the suspension at 35,000 g for 45 min and obtained the supernatant solution (S'2) and the precipitate (P'2). The P'2 fraction was suspended in 6 ml of 20% sucrose. The suspension (2 ml) was layered over a discontinuous sucrose density gradient containing 2 ml of 45% sucrose and 2.5 ml of 30% sucrose. The sucrose density gradient was centrifuged at 6,000 g for 15 min. Three fractions were isolated after centrifugation; fraction A' over the 30% sucrose layer, fraction B' at the boundary between 30% and 45% sucrose and fraction C' at the bottom of the tube. Each fraction was suspended in 10 ml of 0.01 M potassium phosphate buffer (pH 7.0) and centrifuged at 30,000 g for 15 min. A part of the precipitate resuspended in 0.5 ml of the same buffer was examined microscopically. Fraction B' showed a histological property associated with senile plaque, a green birefringence with congo-red staining.

Solubilization of acetylcholinesterase

A protease free, chromatographically pure collagenase was purchased from Advance Biofactures Corporation and pure alkaline protease of streptomyces and trypsin of bovine pancreas (Type XX) were obtained from Toyobo and from Sigma. The fraction B' described above was incubated with collagenase or protease (0.2 mg/ml) for 10 min at 37°C. The incubated mixture (0.05 ml) was diluted with 1.0 ml of cold 0.01 M potassium phosphate buffer, pH 7.0. The diluted mixture was centrifuged at 100,000 g for 60 min and the supernatant solution (S'4) was separated from the precipitate (P'4).

Enzyme assays

AChE und BuChE activities were determined spectrophotometrically using the thiocholine method (Ellman et al., 1961). The reaction was performed at 37°C in a total volume of 4.5 ml and the 5-thio-2-nitrobenzoate produced was measured in absorbancy at a wavelength of 412 nm at 5-min intervals. The reaction proceeded linearly with time up to 20 min.

BuChE activity was determined with butyrylthiocholine as a substrate. Purified bovine BuChE purchased from Sigma chemical company hydrolyzed butyrylthiocholine 1.75 times as fast as acetylthiocholine. Therefore, AChE activity was calculated by the following equation: AChE activity = (hydrolysis rate of acetylthiocholine) − (hydrolysis rate of butyrylthiocholine)/1.75. AChE activity thus obtained coincided well with the value obtained through the use of AChE inhibitor, BW284c51 (Atack et al., 1985), which was donated by the Wellcome Foundation Ltd.

The activity of 5'-nucleotidase and 2'-nucleotidase was measured by methods previously described (Nakamura, 1976; Nakamura et al., 1979). Nonspecific acid phosphatase was assayed according to the method described by Nagata et al. (1984). The modified method of Sottocasa et al. (1967) was employed for the assay of succinic dehydrogenase. Protein was determined by the method of Lowry et al. (1951).

Results

Subcellular distribution of AChE in rat brain cerebral cortex

Less than 10% of AChE activity was observed in the soluble fraction (S4) of the fresh rat cerebral cortex. The highest specific activity of AChE was detected in the nerve ending fraction (B) (Table 1). When 5% TritonX100 $-1M$ NaCl or 2% sodium laurylsulfate was added to the homogenate of rat brain, more than 90% of AChE activity was found in the supernatant solution obtained by centrifugation at 100,000 g for 60 min.

Similar distribution pattern of AChE was observed in the rat cerebral cortex after freezing and thawing. AChE activity in S4 fraction was also less than 10% in autopsied brains either from control subjects or from Alzheimer patients.

Subcellular distribution of AChE in control or Alzheimer brain

While control brains exhibited a similar subcellular distribution pattern to rat brains, AChE activity in Alzheimer brains was recovered mainly in fractions which contained subcellular components with a higher density (P1 or C). The specific activity of AChE was the highest in the fraction C (Fig. 1). Some of structures in the fraction P1 or C were positively stained with congo-red and showed a green birefringence.

AChE activity in senile plaque enriched fraction

The senile plaque enriched fraction was prepared as mentioned in Materials and methods. The highest specific activity of AChE was observed in the

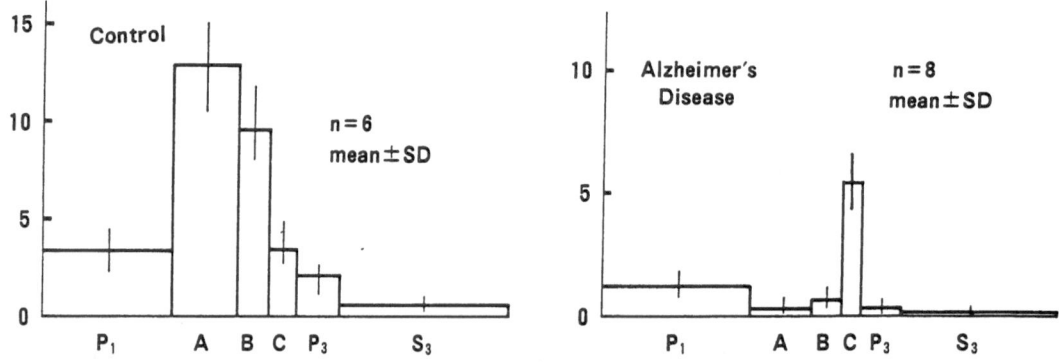

Fig. 1. Subcellular distribution of acetylcholinesterase in human brain. Values are given in nmol/min/mg protein. Vertical lines show the standard deviation. The number of cases is shown in the Figure. The fractionation method and the nomenclature of fractions are described in the text

Table 1. Subcellular distribution of acetylcholinesterase and other enzymes in rat cerebral cortex

| Fractions | Acetylcholinesterase | | | | 5'-Nucleotidase | | 2'-Nucleotidase | | Acid phosphatase | | Succinic dehydrogenase | |
| | Fresh brain | | After freezing | | | | | | | | | |
	Recovery	Specific activity	Recovery	Specific activity	Recovery	Specific activity	Recovery	Specific activity	Recovery	Specific activity	Recovery	Specific activity
Homogenate	100	0.36±0.10	100	0.32±0.12	100	1.48±0.32	100	0.29±0.09	100	1.21±0.35	100	0.86±0.24
P1	18±6	0.37±0.12	16±7	0.34±0.13	17±5	1.32±0.34	6±2	0.09±0.04	12±4	0.78±0.30	19±5	0.85±0.22
A	16±5	0.51±0.17	18±5	0.54±0.15	18±4	2.31±0.42	2±1	0.05±0.02	10±3	1.01±0.38	2±1	0.14±0.08
B	27±7	0.62±0.19	24±8	0.58±0.18	25±6	2.48±0.51	17±4	0.32±0.08	17±5	1.31±0.34	19±4	1.04±0.30
C	11±5	0.28±0.10	11±4	0.24±0.09	6±2	0.73±0.21	6±3	0.14±0.06	12±3	1.20±0.38	52±12	3.65±0.81
P3	11±3	0.20±0.07	12±3	0.22±0.08	25±8	1.84±0.47	3±1	0.04±0.02	25±7	1.54±0.51	4±2	0.28±0.10
S3	10±4	0.16±0.03	12±5	0.14±0.03	3±1	0.21±0.08	63±12	0.85±0.13	20±6	1.11±0.31	1±1	0.04±0.02

Values are given in μmol/h/mg protein. Data are expressed as mean ± SD of 6 experiments. The procedures are described in the text

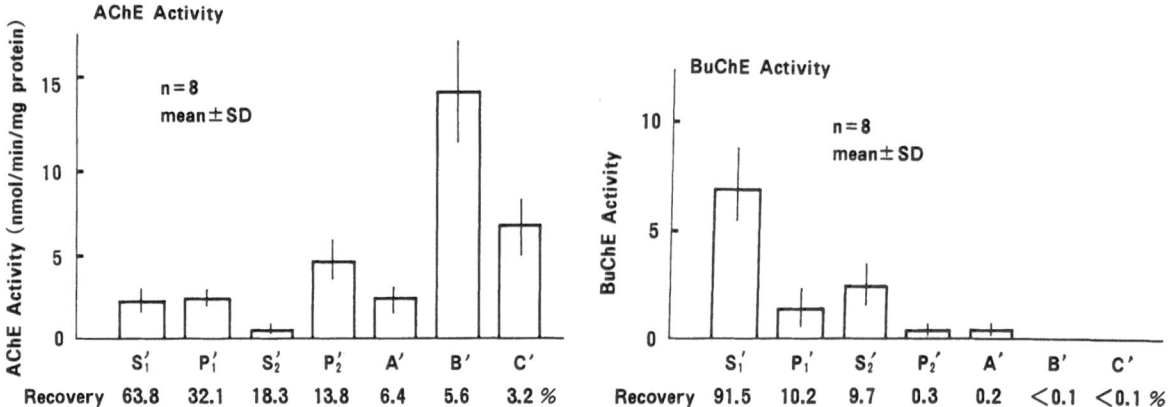

Fig. 2. Activity of AChE and BuChE in fraction enriched with senile plaque. Values are given in nmol/min/mg protein. Vertical lines show the standard deviation. The fraction was obtained from 6 cases with Alzheimer's disease, according to the method described in the text

fraction B′ enriched with senile plaque (Fig. 2). However, the BuChE activity was high in S′1 or S′2, but undetectable in the fraction B′ or C′.

Solubilization of AChE from senile plaque enriched fraction

More than 50% of AChE activity in the fraction B′ appeared in the supernatant solution (S′4) after incubation with either collagenase or protease for 10 min (Fig. 3). A prolonged incubation with collagenase or protease (40 min) resulted a complete solubilization of AChE, although the recovery of AChE activity was less than 60%. Congo-red positive structures were observed in the precipitate (P′4) after the incubation for 40 min.

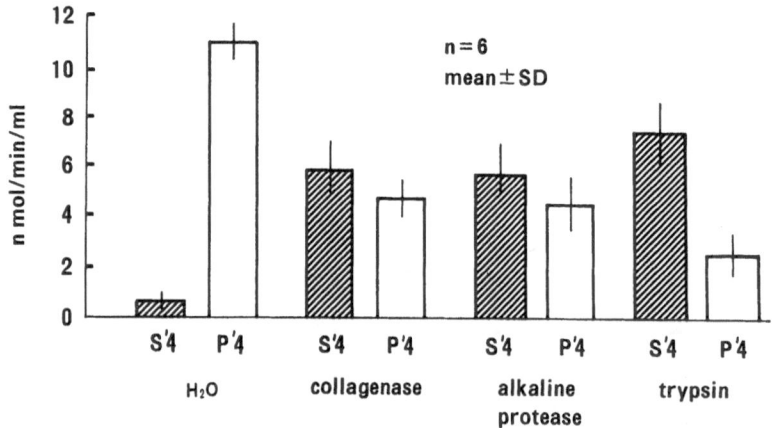

Fig. 3. Solubilization of AChE from fraction enriched with senile plaque. Striped columns show the solubilized AChE activity after the digestion. Values are given in nmol/min/ml. Vertical lines show the standard deviation. The methods are described in the text

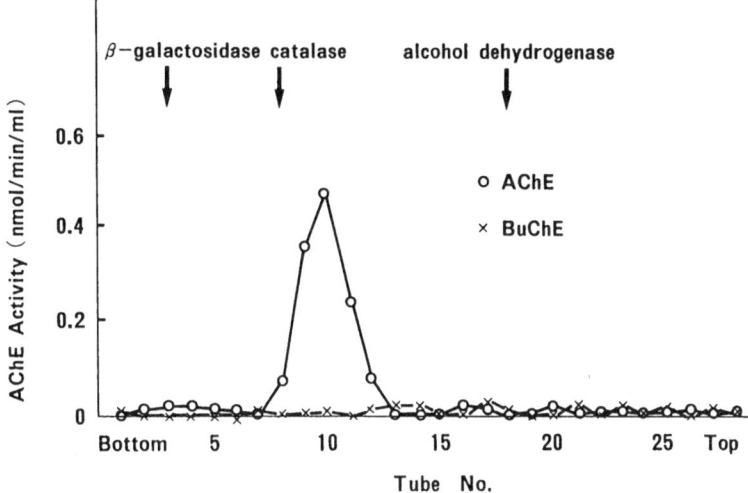

Fig. 4. Sucrose density gradient centrifugation of solubilized preparation. Linear sucrose density gradient (4.5 ml) from 5 to 20% prepared in 20 mM Tris-acetate, pH 8.5. Solubilized preparation (S'4:0.2 ml) was put on the gradient. The centrifugation was performed at 175,000 g for 17.5 h at 4 °C, using catalase, alcohol dehydrogenase and β-galactosidase as standards. Fractions (0.2 ml) were collected from the bottom of the tube. Circles (o) and crosses (×) show AChE and BuChE activity, respectively. The sedimentation coefficient was calculated by the method of Martin and Ames (1961)

Determination of sedimentation coefficient of solubilized AChE

The sedimentation coefficient of AChE in the fraction S'4 was determined by centrifugation on a linear sucrose density gradient. AChE activity was detected at the fraction calculated as 10S, but no AChE activity was found in other fractions (Fig. 4). The activity of BuChE was almost negligible in all fractions separated by the sucrose density gradient centrifugation.

Discussion

The histochemical distribution of AChE has been investigated in brains of patients with Alzheimer's disease and control subjects (Mesulam and Moran, 1987). The cerebral cortex of control subjects exhibited AChE activity mostly within axons and cell bodies belonging to cholinergic pathways. AChE axons were severely depleted in the cerebral cortex of Alzheimer brain. The location of the enzyme was demonstrated by the histochemical method and largely shifted to the senile plaques or neurofibrillary tangles. In the present study neurochemical methods were taken to investigate the nature of AChE and the mechanism of the abnormal localization.

As reported previously (Whittaker et al., 1964; McIntosh and Plummer, 1974), high specific activity of AChE was found in the nerve ending fraction. The distribution pattern of AChE was different from those of other enzymes which have been known to localize in certain subcellular components: 5'-nucleotidase (plasma membranes), 2'-nucleotidase (soluble fraction), acid phosphatase (lysosomes) and succinic dehydrogenase (mitochondria).

However, results obtained by the subcellular fractionation procedure should be interpreted with caution (Brzin et al., 1983), especially after freezing and thawing, since these procedures would disrupt particles such as mitochondria or lysosomes. Actually, electronmicroscopical examinations revealed a membrane disruption and the subcellular distribution of most enzymes listed above showed a critical change after freezing and thawing. But little difference was noted in AChE distribution pattern between fresh and frozen rat brains. The result has encouraged us to study on the subcellular distribution of AChE in the autopsied human brain. Most of the information on cortical AChE has been obtained from laboratory animals, using histochemical methods. Less details are available on AChE of the human cerebral cortex by neurochemical methods.

Although the essential difference was not recognized between the rat brain and the control human brain, the distribution of AChE activity in Alzheimer brain was shifted to heavier components found in fraction P1 or C containing amyloid protein. Various procedures have been developed to isolate senile plaque core (Allsop et al., 1983; Candy et al., 1986). A considerable amount of AChE activity in Alzheimer brain (13.8%) was recovered in the precipitate after solubilization, whereas most part of AChE activity in control brain (>98%) was detected in the supernatant solution by the same procedure. The particulate fraction (B') enriched with senile plaque showed the highest AChE specific activity. These biochemical results confirmed the histochemical observation that AChE is localized in senile plaque (Mesulam and Moran, 1987).

The activity of BuChE was found mainly in the supernatant solution after sonication and treatment with detergent and practically no activity was observed in the senile plaque rich fraction (B'). The present result seems to contradict the histochemical finding that BuChE is abundant in senile plaque and tangles (Mesulam and Moran, 1987). Probably the discrepancy may derive from the different mechanism of attachment to senile plaque between AChE and BuChE, the latter of which seems to be readily solubilized by sonication or with detergents.

AChE was solubilized from isolated B' fraction after incubation either with protease or with collagenase, while amyloid protein was left precipitated as has been reported (Allsop et al., 1983; Candy et al., 1986). Solubilized AChE showed the sedimentation coefficient 10S which corresponded with that of G4 isozyme. Solubilization with protease-free collagenase suggests that AChE in the senile plaque, probably amyloid core would be an A form possessing a collagen tail (Massoulie and Bon, 1982). The globular G4 isozyme might be detached from the senile plaque by digestion with collagenase. Further confirmation on the molecular form of AChE in the senile plaque is under investigation using the sequential extraction method with different concentrations of salt developed by Younkin et al. (1982).

AChE has been reported to show also a trypsin-like activity in addition to its well-known esterase activity, and collagen tail is supposed to be cleaved by autolysis (Small and Simpson, 1988). Recently, Kitaguchi et al. (1988) have reported a gene coding a protease inhibitor near the locus of amyloid protein in the chromosome 21. The messenger RNA for the protease inhibitor seems increased in the Alzheimer brain (Tanaka et al., 1988). The presence of α1-antichymotrypsin has also been demonstrated in the Alzheimer brain (Abraham et al., 1988). The protease inhibitor might prevent the conversion from an asymmetric form to a globular form of AChE as well as the break down of amyloid protein.

The effect of AChE inhibitor has been studied in autopsied human brains by a histochemical method (Mesulam et al., 1987; Geula and Mesulam, 1988). But the precise quantitative examination may be possible only through neurochemical procedures. Further neurochemical studies will evaluate the utility of AChE inhibitor for the therapy of Alzheimer's disease.

Acknowledgments

This work was supported by a grant from the special project research of molecular biology of aging brain by the Ministry of Education, Science and Culture of Japan. The authors wish to thank Drs. M. Ogawa, Hamamatsu Rosai Hospital and M. Kato, Biwako Yoikuin Hospital for collecting the autopsied brains.

References

Abraham CR, Selkoe DJ, Potter H (1988) Immunochemical identification of the serine protease inhibitor α1-antichymotrypsin in the brain amyloid deposits of Alzheimer's disease. Cell 52:487–501

Allsop D, Landon M, Kidd M (1983) The isolation and amino acid composition of senile plaque core protein. Brain Res 259:348–352

Atack JR, Perry EK, Perry RH, Wilson ID, Bober MJ, Blessed G, Tomlinson BE (1985) Blood acetyl- and butyrylcholinesterase in senile dementia of Alzheimer type. J Neurol Sci 70:1–12

Brzin M, Sketelj J, Klinar B (1983) Cholinesterases. In: Lajtha A (ed) Handbook of neurochemistry, vol 4. Plenum Press, New York, pp 251–292

Candy JM, Klinowski J, Perry RH, Perry EK, Fairbairn A, Oakley AE, Carpenter TA, Atack JR, Blessed G, Edwardson JA (1986) Aluminosilicates and senile plaque formation in Alzheimer's disease. Lancet i:354–357

Davies P, Maloney AJF (1976) Selective loss of central cholinergic neurons in Alzheimer's disease. Lancet ii:1403

Ellman GL, Courtney KD, Andrews V Jr, Featherstone RM (1961) A new and rapid colorimetric determination of acetylcholinesterase activity. Biochem Pharmacol 7:88–95

Friede RL (1965) Enzyme histochemical studies of senile plaques. Neuropathol Exp Neurol 24:477–491

Geula C, Mesulam M (1988) Enzymatic properties of cholinesterases in normal human brain and Alzheimer's disease. Soc Neurosci Abstr 14:155

Gray EG, Whittaker VP (1962) The isolation of nerve endings from brain. An electronmicroscopic study of all fragments derived by homogenisation and centrifugation. J Anat 96:79–88

Hammond P, Brimijoin S (1988) Acetylcholinesterase in Huntington's and Alzheimer's diseases: simultaneous enzyme assay and immunoassay of multiple brain regions. J Neurochem 50:1111–1116

Kitaguchi N, Takahashi Y, Tokushima Y, Shiojiri S, Ito H (1988) Novel precursor of Alzheimer's disease amyloid protein shows protease inhibitory activity. Nature 331:530–532

Koshimura K, Kato T, Tohyama I, Nakamura S, Kameyama M (1986) Qualitative abnormalities of choline acetyltransferase in Alzheimer type dementia. J Neurol Sci 76:143–150

Lowry OH, Rosebrough NJ, Farr AL, Randall RJ (1951) Protein measurement with the Folin phenol reagent. J Biol Chem 193:265–275

Martin RG, Ames BN (1961) A method for determining the sedimentation behavior of enzymes – Application to protein mixtures. J Biol Chem 236:1372–1379

Massoulie J, Bon S (1982) The molecular forms of cholinesterase and acetylcholinesterase in vertebrates. Ann Rev Neurosci 5:57–106

McIntosh CHS, Plummer DT (1976) The subcellular localization of acetylcholinesterase and its molecular forms in pig cerebral cortex. J Neurochem 27:449–457

Mesulam MM, Geula C, Moran MA (1987) Anatomy of cholinesterase inhibition in Alzheimer's disease: effect of physostigmine and tetrahydroaminoacridine on plaques and tangles. Ann Neurol 22:683–691

Mesulam MM, Moran MA (1987) Cholinesterases within neurofibrillary tangles related to age and Alzheimer's disease. Ann Neurol 22:223–228

Nagata H, Mimori Y, Nakamura S, Kameyama M (1984) Regional and subcellular distribution in mammalian brain of the enzymes producing adenosine. J Neurochem 42:1001–1007

Nakamura S (1976) Effect of sodium deoxycholate on 5′-nucleotidase. Biochim Biophys Acta 426:339–347

Nakamura S, Yamao S, Ito J, Kameyama M (1979) Purification and properties of 2′-nucleotidase from mammalian brain. Biochim Biophys Acta 568:30–38

Nakano S, Kato T, Nakamura S, Kameyama M (1986) Acetylcholinesterase activity in cerebrospinal fluid of patients with Alzheimer's disease and senile dementia. J Neurol Sci 75:213–223

Perry EK (1980) The cholinergic system in old age and Alzheimer's disease. Age Ageing 9:1–8

Perry EK, Gibson PH, Blessed G, Perry RH, Tomlinson BE (1977) Neurotransmitter enzyme abnormalities in senile dementia. Choline acetyltransferase and glutamic acid decarboxylase activities in necropsy tissue. J Neurol Sci 34:247–265

Reisberg B (1986) Dementia: a systematic approach to identifying reversible causes. Geriatrics 41(4):30–46

Small DH, Simpson RJ (1988) Acetylcholinesterase undergoes autolysis to generate trypsin-like activity. Neurosci Lett 89:223–228

Sottocasa GL, Kuylenstierna B, Ernster L, Bergstrand A (1967) An electron-transport system associated with the outer membrane of liver mitochondria. J Cell Biol 32:415–438

Tanaka S, Nakamura S, Ueda K, Kameyama M, Shiojiri S, Takahashi Y, Kitaguchi N, Ito H (1988) Three types of amyloid precursor mRNA in human brain: their differential expression in Alzheimer's disease. Biochem Biophys Res Commun 157:472–479

Whittaker VP, Michaelsson IA, Kirkland JA (1964) The separation of synaptic vesicles from nerve-ending particles (synaptosomes). Biochem J 90:293–303

Younkin SG, Goodridge B, Katz J, Lockett G, Nafziger D, Usik MF, Younkin LH (1986) Molecular forms of acetylcholinesterase in Alzheimer's disease. Fed Proc 45:2982–2988

Younkin SG, Rosenstein C, Collins PL, Rosenberry TL (1982) Cellular localization of the molecular forms of acetylcholinesterase in rat diaphragm. J Biol Chem 257:13630–13637

Authors' address: Dr. S. Nakamura, Department of Neurology, Kyoto University Hospital, Shogoin Sakyo-ku, Kyoto 606, Japan.

J Neural Transm (1990) [Suppl] 30: 25–32

Changes of acetylcholine and choline concentrations in cerebrospinal fluids of normal subjects and patients with dementia of Alzheimer-type

Y. Ikeda, S. Okuyama, Y. Fujiki, K. Tomoda, K. Ohshiro, T. Itoh, and **T. Yamauchi**

Department of Psychiatry, School of Medicine, Fujita Health University, Aichi, Japan

Summary. Acetylcholine (ACh) and choline (Ch) in cerebrospinal fluid from 29 normal volunteers and 7 patients with Alzheimer-type dementia (DAT) were examined using high-performance liquid chromatography with electrochemical detector coupled with liquid cation-exchange method. In normal volunteers, ACh concentration was decreased significantly from 40–50 years and Ch concentration was increased significantly from 50–60 years. CSF from patients with DAT revealed high Ch concentration and the increase was statistically significant while ACh concentration in CSF of DAT did not show a significant difference with that of normal volunteers. This Ch augmentation may suggest a disturbance in utilization of Ch for ACh synthesis and may become an useful indicator for organic changes in central cholinergic system.

Introduction

Dementia of Alzheimer-type (DAT) is characterized by the degeneration of neurons in the central cholinergic system and the appearance of neurofibrillary tangles and senile plaques in the brain, which appeared partially in normal aged people. Several biochemical changes in the central nervous system of patient with DAT have been pointed out including reduced activity of choline acetyltransferase (ChAT) and acetylcholinesterase (ChE), indicating reduced activity in the cholinergic system in the central nervous system.

In 1983, Potter et al. first developed an assay for ACh and Ch by high-performance liquid chromatography with electrochemical detection (HPLC-ED) in brain tissues. Further, an immobilized enzyme reactor for assaying ACh instead of enzyme infusion (Potter et al., 1983) was developed (Fujimori and Yamamoto, 1987). This method is widely used for measuring the concentration not only in the brain but in many other tissues and also in perfusate obtained from various tissues. It is difficult, however, to use this system directly for measuring ACh concentrations in cerebrospinal fluid (CSF) because CSF contains a very low concentration of ACh. The liquid-cation exchange method is a procedure for extracting choline compounds. By using this procedure coupled with the HPLC-ED system, we were able to measure ACh and Ch contents simultaneously in human CSF (Okuyama and Ikeda, 1988).

In this paper we described an ageing-induced change of ACh and Ch concentrations in human CSF, and also changes of these substances in the CSF of patients with DAT.

Materials and methods

Physostigmine and tetraphenylboron were obtained from Sigma (St. Louis, MO, U.S.A.). Sodium 1-decanesulphonate, choline chloride, acetylcholine chloride and 3-heptanone were obtained from Tokyo Kasei (Tokyo, Japan). Tetramethylammonium chloride, disodium hydrogenphosphate, sodium dihydrogenphosphate and other chemicals were obtained commercially. Ethylhomocholine (EHC), which was used as an internal standard, was synthesized from 3-dimethylamino-1-propanol and iodoethane (Potter et al., 1983).

Samples of human CSF were collected from 29 volunteers free from neurological and psychological diseases. The volunteers, nine males and sixteen females, were from 20 to 78 years old (mean $\pm$ S.D., 51.5 ± 18.1 years). The subjects were divided into five groups according to their age. Seven patients affected with DAT (five females and two males, mean age 62.7 ± 11.2 years) were examined. The diagnosis of DAT was made by clinical criteria, according to DSM III-R. They had not any signs and symptoms of vascular disorders. Lumbar puncture was performed before medication.

We collected 2 ml of human CSF by lumbar puncture in a plastic tube containing 2 μmol of eserine and kept it at 4°C until extraction.

Extraction of ACh and Ch from CSF was performed within 12 h after collection according to the liquid cation-exchange method (Fonnum, 1969) as shown in Fig. 1. In brief, the samples were shaken with 1 ml of 3-heptanone containing 10 mg of tetraphenylboron for 10 min at room temperature and centrifuged at 25,000 g for 20 min at 4°C. Then 500 μl of the supernatant were taken and added to 0.5 ml of 0.4 M hydrochloric acid. The mixture was shaken for 1 min and centrifuged at 1,500 g for 5 min at 4°C. Finally, the organic layer was discarded by aspiration and the acid layer collected, lyophilized and stored at 4°C. Samples were dissolved in 35 μl of distilled water and 20 μl were injected into the HPLC system.

The assay method is based on the separation of ACh and Ch on a Shodex RSpak column, followed by their enzymatic conversion to hydrogen peroxide through a post-column immobilized enzyme reaction with acetylcholinesterase and choline oxidase (Potter et al., 1983). ACh, Ch and EHC could be measured with a high sensitivity using the HPLC-ED system.

The HPLC system consisted of an 880-50 degasser (Japan Spectroscopic, Tokyo, Japan), an EP-10 pump, a CB-100 amperometric detector equipped with a WE-PT platinum electrode (Eicom, Kyoto, Japan), a Rheodyne 7125 injector with a 200 μl sample loop (Berkeley, CA, U.S.A.), a Prepak guard column (5 × 4 mm I.D.) (Eicom),

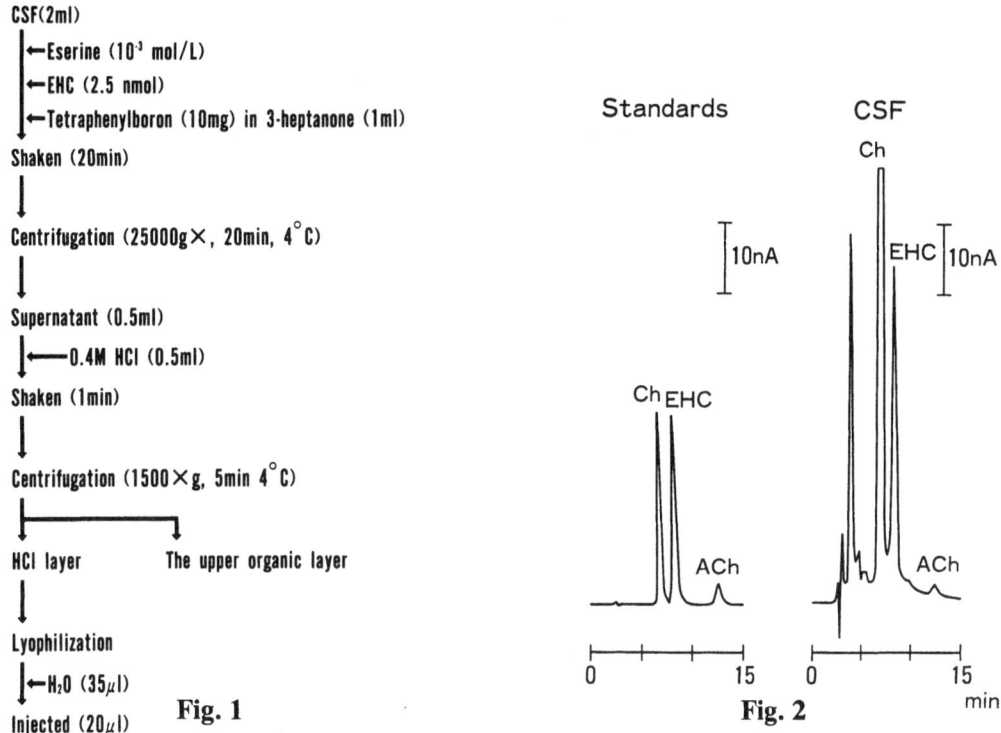

Fig. 1. Extraction of ACh and Ch from CSF

Fig. 2. Chromatograms of standard samples and a human CSF extract containing 2.5 nmol of EHC as an internal standard. The standard samples contained 15.6 pmol of ACh and 156 pmol each of Ch and EHC

an RSpak DE613 (medium polarity methacrylate gel, 150×6 mm I.D.) (Shodex, Tokyo, Japan), an AC-Enzympak immobilized enzyme column (5×4 mm I.D.) (Eicom). The mobile phase was 0.1 M sodium phosphate buffer (pH 8.3) containing 1.2 mM tetramethylammonium chloride (TMA) and 300 mg/l sodium 1-decanesulphonate, which was filtered through a 0.22 μm membrane filter (Millipore, Bedford, MA, U.S.A.). The HPLC separation and enzymatic reaction were performed at 37°C. The flow-rate was 1.0 ml/min. The electrode potential was set at $+450$ mV against an Ag/AgCl reference electrode for the detection of hydrogen peroxide. Under these conditions the retention times were Ch, 6.8 min, EHC, 8.3 min and ACh, 12.8 min.

The peak areas increased linearly with increasing volume injected for ACh from 0.3 pmol to 5 nmol and for Ch from 1 pmol to 5 nmol. The calibration graph (n = 3) for ACh showed linearity in the range 1–60 pmol and that for Ch in the range 10–300 pmol. The recoveries (n = 3) of added ACh (312 pmol per 2 ml), Ch (3.1 nmol per 2 ml) and EHC (3.1 nmol per 2 ml) were ACh 91.7%, Ch 104.5% and EHC 88.5%.

Figure 2 shows typical chromatograms of a human CSF sample (containing 156 pmol of EHC as an internal standard) obtained from a 35-year-old volunteer and of authentic standard samples (156 pmol of Ch, 156 pmol of EHC and 15.6 pmol of ACh). The peaks of Ch, EHC and ACh from the CSF sample were clearly separated and were identified by comparison with the authentic samples.

A statistical analysis was performed by the Wilcoxon rank-sum test.

Results

Concentrations of ACh in CSF from normal subjects (n = 29) ranged from 0 to 268.9 pmol/ml. The mean and standard deviation for all the specimens were 81.2 ± 65.8 pmol/ml. Contents of ACh decreased with advancing age (Fig. 3). The third and fourth decade showed higher concentrations. Markedly low concentrations of CSF ACh were demonstrated in subjects older than 40 years old. This variation was statistically significant according to the Wilcoxon rank-sum test.

The CSF concentrations of Ch in 29 normal human subjects ranged from 684.1 to 4338.0 pmol/ml. Mean and SD were 2480.2 and 981.3 pmol/ml, respectively. The Ch concentration increased with the age (Fig. 4). There was a significant difference in Ch concentration in aged, compared with third decade (Fig. 4).

We calculated a value that was obtained by deviding ACh concentration with sum of ACh and Ch concentrations. This index also decreased with the age (Fig. 5).

The Ch concentrations of specimen obtained in later fractions, tend to be higher than that in the first fraction. There was a statistical significance between concentration of Ch in the first and the last fractions (Fig. 6).

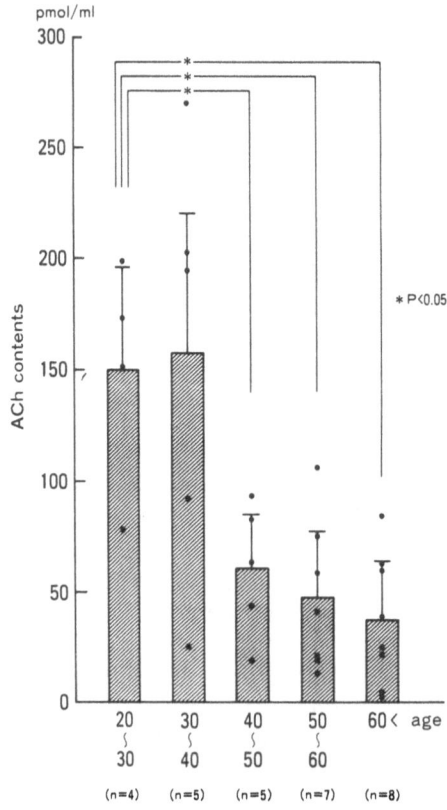

Fig. 3. Contents of ACh in normal human CSF (mean ± S.D.); * p < 0.05

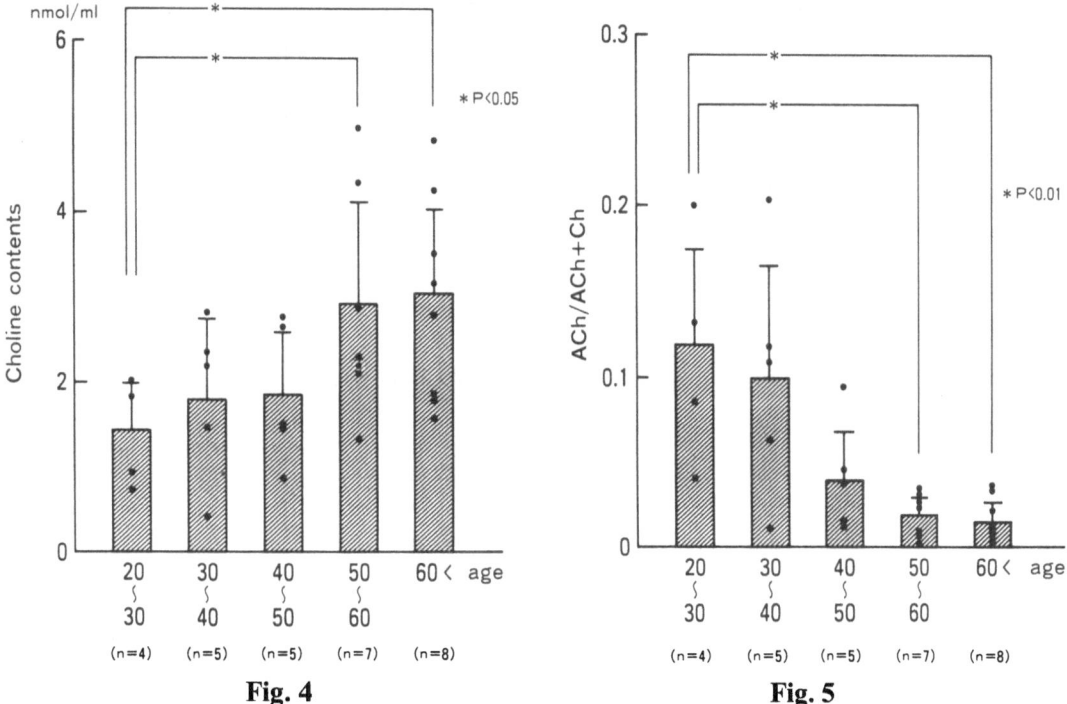

Fig. 4. Contents of Ch in normal human CSF (mean ± S.D.); * p < 0.05

Fig. 5. Change of $\left(\dfrac{\text{ACh contents}}{\text{ACh and Ch contents}}\right)$ as a function of age

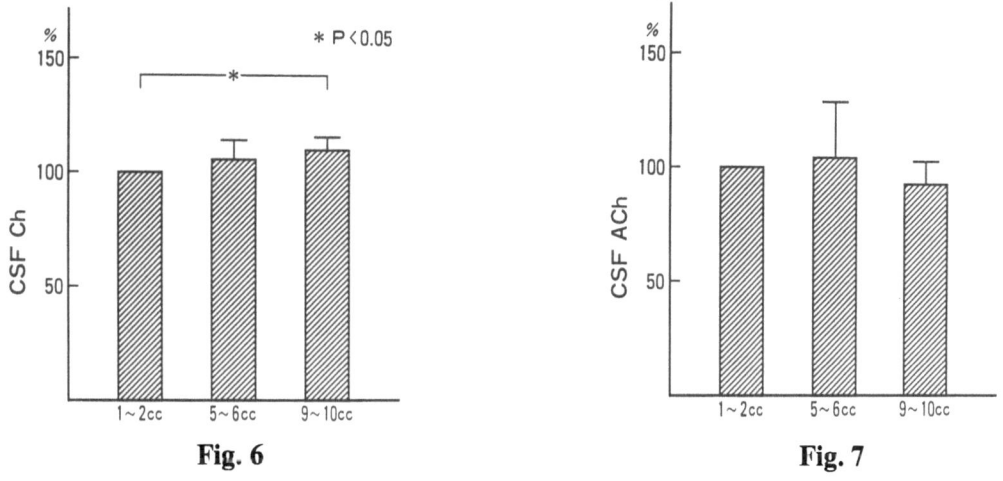

Fig. 6. A gradient of Ch concentration in normal human CSF

Fig. 7. A gradient of ACh concentration in normal human CSF

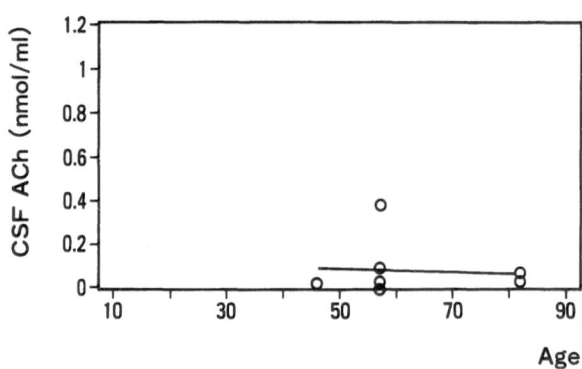

Fig. 8. ACh contents in CSF from patients with DAT

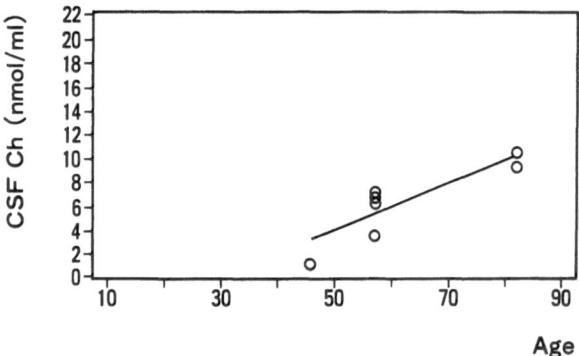

Fig. 9. Ch contents in CSF from patients with DAT

However, we could not find any significant change in ACh concentration between the first and the last fraction (Fig. 7). Ch concentration in CSF may reflect the cholinergic disturbance in the brain rather than ACh concentration.

ACh concentration in the CSF of DAT ranged from 0 to 383.0 pmol/ml. Mean value was 84.4 pmol/ml. Markedly a high ACh concentration was observed in one case and the other cases had no significant difference from normal subjects.

Ch concentration in the CSF of DAT ranged from 1.82 to 10.54 nmol/ml. Mean and S.D. were 6.44 and 2.79 nmol/ml, respectively. Ch concentration in the CSF of DAT was high, compared with that of normal subjects. This difference was statistically significant ($p < 0.05$).

Discussion

Few reports were published on ageing-induced change of ACh concentration in normal human CSF. Recently we demonstrated the ageing-induced reduction of ACh concentration in human CSF (Okuyama and Ikeda, 1988). In this work, we presented a more precise examination about ageing-induced changes of ACh and Ch concentrations in normal human CSF and further demonstrated increased concentration of Ch in CSF obtained from patients with DAT.

Previously, ChAT and ChE activities in autopsied brains or CSF were measured by many investigators for an evaluation of central cholinergic activities (McGeer and McGeer, 1976). The reduction of ACh concentration in middle-aged and elderly people described in this paper is compatible with the McGeers' findings that indicated the ageing-induced reduction of cholinergic activity in the central nervous system, especially the decrease in ChAT activity.

Ch concentrations of CSF from normal subjects tend to increase following the advancing age. The increase in Ch concentration in 50–70 years was statistically significant compared with that in 20–29 years. ACh is produced from Ch and acetyl CoA by ChAT and degraded to Ch and acetic acid by ChE. So, Ch is the metabolite of ACh and also the precursor of ACh. We calculated a value by dividing ACh concentration with sum of ACh and Ch concentrations. This index decreased following the advancing age. These results may suggest a decline of ACh-Ch turnover in the central nervous system of aged people.

The CSF ACh has been reported to be originated from the caudate nucleus (Beleslin et al., 1964). So, ageing-induced decrease of ACh in CSF might reflect a change of cholinergic activity in the caudate nucleus. However, ACh concentration in CSF and memory function would decrease simultaneously in DAT without any other changes in neurotransmitters (Davis et al., 1982).

We have already developed a new memory test using a meaningful and meaningless syllables (Ikeda et al., 1989) and investigated the memory function of normal subjects. Memory function decreased gradually according to their age using the test of meangingful syllables. Memory of meaningless syllables was significantly decreased after 40 years old. The attention has been paid to the close relation between decrease of ACh concentration and decline of memory function. Further examination would be necessary on the issue.

The concentration of CSF ACh obtained from patients with DAT in this work was in agreement with previous published data (Davis et al., 1982, 1985). However, Ch concentration in the CSF of patients with DAT was relatively high in our study, compared with other published data (Davis et al., 1982). Ch concentration in the CSF of patients with Parkinson's disease was increased, compared with normal control (Welch et al., 1976). It is of interest that the degenerative process of CNS causes an increase in Ch concentration. Ch augumentation may be resulted from the disturbance of utilization of Ch for ACh synthesis.

References

Beleslin D, Carmichale EA, Feldberg W (1964) The origin of acetylcholine appearing in the effluent of perfused cerebral ventricles of the cat. J Physiol 173:368–376

Davis KL, Hsieh JYK, Levy MI, Horvath TB, Davis BM, Mohs RC (1982) Cerebrospinal fluid acetylcholine, choline and senile dementia of the Alzheimer's type. Psychopharmacol Bull 18:193–195

Davis BM, Mohs RC, Greenwald BS, Mathe AA, Johns CA, Horvath TB, Davis KL (1985) Clinical studies of the cholinergic deficit in Alzheimer's disease. J Am Geriatr Soc 33:741–748

Fonnum F (1969) Radiochemical microassays for the determination of choline acetyl-transferase and acetylcholinesterase activities. Biochem J 115:465–472

Fujimori K, Yamamoto K (1987) Determination of acetylcholine and choline in perchlorate extracts of brain tissue using liquid chromatography – electrochemistry with an immobilized-enzyme reactor. J Chromatogr 414:167–173

Ikeda Y, Ohshiro K, Tomoda K, Fujiki Y, Okuyama S, Itoh T, Yamauchi T (1989) Ageing-induced memory deficit indicated by newly-developed memory test. Abstr Fourth Congress of the International Psychogeriatric Association, p 120

McGeer EG, McGeer PL (1976) Neurotransmitter metabolism in the aging brain. In: Terry RD, et al (eds) Neurobiology of aging. Raven Press, New York, p 389

Okuyama S, Ikeda Y (1988) Determination of acetylcholine and choline in human cerebrospinal fluid using high-performance liquid chromatography combined with an immobilized enzyme reactor: ageing-induced change of acetylcholine level. J Chromatogr 431:389–394

Potter PE, Meek JL, Neff NHJ (1983) Acetylcholine and choline in neuronal tissue measured by HPLC with electrochemical detection. J Neurochem 41:188–194

Welch MJ, Markham CH, Jenden DJ (1976) Acetylcholine and choline in cerebrospinal fluid of patients with Parkinson's disease and Huntington's chorea. J Neurol Neurosurg Psychiatry 39:367–374

Authors' address: Dr. Y. Ikeda, Department of Psychiatry, School of Medicine, Fujita Health University, 1-98 Dengakugakubo, Kutsukake, Toyoake, 470-11 Aichi, Japan.

J Neural Transm (1990) [Suppl] 30: 33–43

Disturbance of the 5-hydroxytryptamine metabolism in brains from patients with Alzheimer's dementia

C. G. Gottfries

Department of Psychiatry and Neurochemistry, Gothenburg University,
Gothenburg, Sweden

Summary. The 5-hydroxytryptamine (5-HT) system in the human brain is sensitive to aging. In dementia of the Alzheimer type (AD/SDAT), there are significantly reduced concentrations of 5-HT and 5-hydroxyindole-acetic acid (5-HIAA). 5-HT-sensitive imipramine binding is reduced by almost 50%, indicating a loss of presynaptic 5-HT terminals. There also seems to be reduced tryptophan hydroxylase activity in some brain areas. In cerebrospinal fluid (CSF) from AD/SDAT patients, the concentration of 5-HIAA is reduced, and the accumulation of 5-HIAA after probenecid loading is diminished. Biochemical findings together with structural findings in the raphe nuclei indicate that the disturbance of the 5-HT system is of the same magnitude as the disturbance of the cholinergic system.

Reduced activity in the 5-HT system may be of importance for activity in the hypothalamus. There is an increased concentration of arginine vasopressin, which may explain the increased activity in the hypothalamic-pituitary-adrenal axis seen in patients with AD/SDAT. This activity is reduced when a selective 5-HT reuptake blocker is given.

Pharmacological treatment with 5-HT reuptake blockers improves emotional disturbances, confusion, anxiety and depressed mood in patients with AD/SDAT.

Introduction

Serotonin (5-HT)-containing neurons and fibres have been visualized in the various raphe nuclei of the brain stem. The nigrostriatal system is exclusively innervated from the dorsal raphe nucleus, whereas cortical regions and limbic structures receive 5-HT projections from the dorsal, medial, and mesencephalic raphe nuclei.

The more selective 5-HT reuptake blockers lately introduced have created a renaissance for the 5-HT system. Other drugs have also been developed that have an agonistic or antagonistic effect on 5-HT receptors. As is well known, there seem to be various subtypes of 5-HT receptors. At present, intense research is being conducted on classification of the multiple serotonin receptors, using radio ligand-binding techniques. A major problem, however, when analysing the functional role of 5-HT, is that some 5-HT receptors exert effects opposite to others. The specific role of 5-HT is therefore incompletely known. The availability of new pharmacological tools would lead to a better understanding of the 5-HT system in the brain.

Animal data indicate that arousal and sleep mechanisms are to some extent influenced by the 5-HT system. Endocrine regulation also seems to be controlled by the 5-HT system. A central clock and the control of hypothalamic and hypophyseal hormones seem to be dependent on the 5-HT system. Sympathetic nervous activity, such as blood pressure and temperature regulation, seems to be regulated by 5-HT neurons.

Crude 5-HT manipulations have shown that the function of the 5-HT system in the mammalian brain also influences more integrated and complex behaviour. Feeding, reproductive, affective and defensive function, anxiety, and aggressiveness are thus in some way related to the function of the 5-HT system.

It has long been agreed that disturbance of the 5-HT system may be part of the pathophysiology of affective disorders, panic, anxiety, feeding disorders, and uncontrolled aggressiveness.

To this list can now dementia disorders be added. Data supporting the hypothesis of a disturbance of the 5-HT system in patients with dementia of the Alzheimer type (AD/SDAT) are reviewed below. However, it has also been shown that the 5-HT system is severely disturbed in vascular dementia (Carlsson and Gottfries, 1986; Wallin et al., 1989).

Disturbance of the 5-HT system in AD/SDAT

Postmortem investigations

Figure 1 shows the metabolism of 5-HT. Most of the data on 5-HT metabolism in the human brain come from postmortem investigations. In these studies, the concentrations of 5-HT and 5-hydroxyindoleacetic acid (5-HIAA) are measured in discrete areas of the brain. The 5-HT concentration is assumed to be mainly a marker of the number of 5-HT neurons or terminals, while the concentration of 5-HIAA is assumed to be a marker also of metabolic activity in these neurons. Usually the precursors tryptophan and 5-hydroxytryptophan are not measured, as changes after death in these amino acids make the interpretation of results difficult.

In postmortem investigations of the globus pallidus, hippocampal cortex (Bucht et al., 1981), caudate nucleus, putamen and hypothalamus

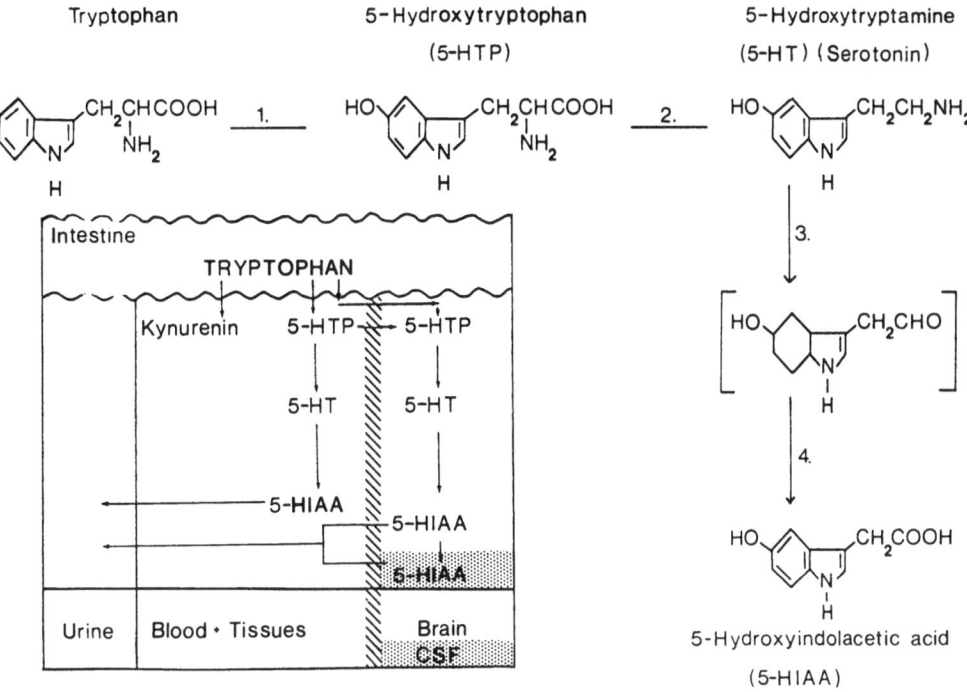

Fig. 1. The metabolism of serotonin

(Carlsson and Gottfries, 1986; Gottfries et al., 1986), 5-HT has been found to be significantly reduced with age. The patient sample in these investigations has been 60 years or older. There are also investigations, however, in which no significant reduction of 5-HT with age has been found, for instance, in the frontal cortex and in the basal ganglia (Carlsson et al., 1980; Bucht et al., 1981). In one study, there was a significant positive correlation between the concentration of 5-HT in the brain stem and age (Carlsson et al., 1980).

It is of interest that while 5-HT, at least in cortical areas and basal ganglia, is significantly negatively correlated with age, the end metabolite 5-HIAA is not. In studies at our institute (Gottfries et al., 1983), 5-HIAA has continually been found not to be correlated with age and, to our knowledge, no data on a correlation between 5-HIAA and age have been published. This may indicate that, although the number of 5-HT neurons is reduced with age, as reflected in reduced 5-HT concentrations, the loss of neurons is compensated for by an increased turnover rate of the remaining neurons, as reflected in the unchanged 5-HIAA concentration. This is possibly attributable to a compensatory mechanism of the neurons (Carlsson, 1986).

Although no significant correlations have been found between age and brain tissue 5-HIAA concentrations, there is a significant positive correlation between cerebrospinal fluid (CSF) 5-HIAA levels and age (Gottfries

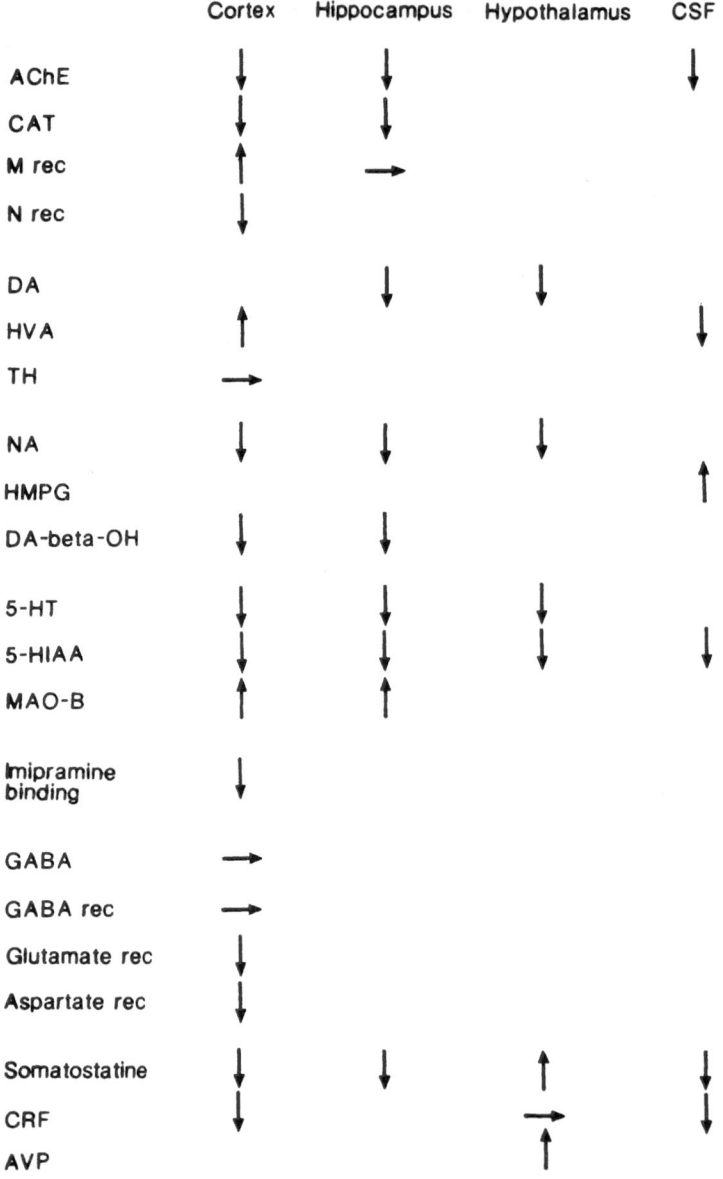

Fig. 2. Changes in neurotransmitters in patients with Alzheimer-type dementia

et al., 1971). There is as yet no answer to the question whether this increase of 5-HIAA in CSF is caused by an increased release of 5-HT in the brain or the spinal cord, or by a reduced discharge of 5-HIAA from the liquor space.

There is evidence from several studies on dementia of the Alzheimer type that the concentrations of 5-HT and 5-HIAA are reduced in brains from demented patients compared with age-matched controls (Gottfries et al., 1976; Arai et al., 1984; for a review see Hardy et al., 1985; Gottfries, 1988). The reduced concentrations of 5-HT and 5-HIAA are found not

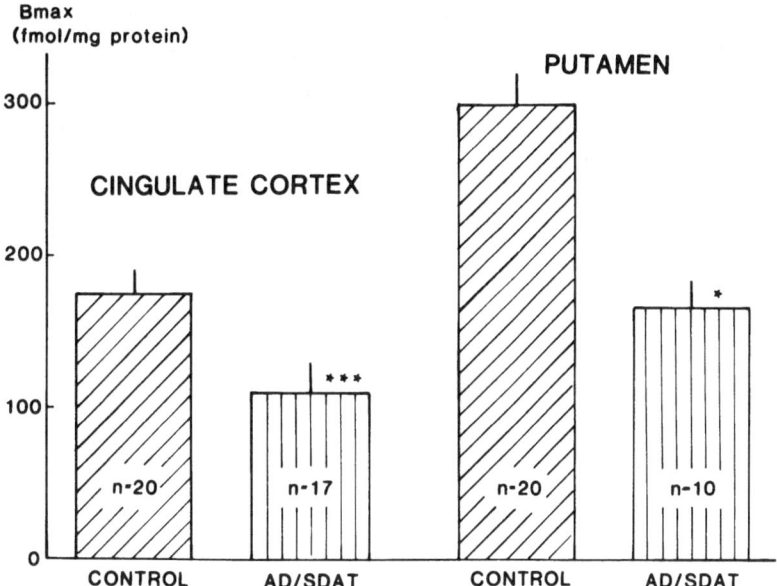

Fig. 3. Imipramine binding in postmortem human brain tissue

only in cortical areas, but also in basal ganglia and brain stem areas (Fig. 2).

The 5-HT-sensitive imipramine binding in postmortem human brain tissue is considered a marker of the presynaptic 5-HT neuron. A significant decrease in B_{max} with age was found in the cortex gyrus cinguli, while in the frontal cortex, hippocampus, amygdalae and putamen no such correlation was found (Marcusson et al., 1987).

The 5-HT-sensitive imipramine binding in tissue from AD/SDAT brains shows an almost 50% decrease in B_{max} when compared with age-matched controls (Marcusson et al., 1987) (Fig. 3).

The synthesizing enzyme of 5-HT, tryptophan hydroxylase (TPH), has been investigated by Nagatsu and Iizuka (1989). TPH activity was found to be lower in various brain regions, and the differences were significant in the lateral segments of the globus pallidus, in the locus caeruleus and the substantia nigra.

Mann and Yates (1983) showed degenerative changes in the raphe nucleus in patients with AD/SDAT.

Monoamine oxidase (MAO) is the main metabolizing enzyme of 5-HT. It exists in two forms, A and B. 5-HT is mainly the substrate for the A-form. In normal aging, there is a significant increase in MAO-B. This increase is seen in both grey and white matter (Adolfsson et al., 1981; Oreland and Gottfries, 1986). Results from animal experiments indicate that increased MAO-B activity may be attributable to an increase in extra-neuronal tissue in the aged brain. Thus, MAO-B seems to be a marker of gliosis. It is of interest that MAO in platelets also increases with age

(Adolfsson et al., 1981). MAO-B is considerably more increased in patients with AD/SDAT than in age-matched controls in grey as well as white matter (Figs. 4, 5). This may indicate that there is a more pronounced gliosis in the demented brain than in the aged brain. The biological importance of increased MAO-B activity is difficult to speculate upon. As 5-HT is mainly the substrate of the A-form, it can be assumed that the increase in the B-form is of no biological importance for the 5-HT system. In platelets, there is only the MAO-B form, and increased platelet MAO activity has been recorded in patients with AD/SDAT compared with age-matched controls (Oreland and Gottfries, 1986). This increase has been difficult to explain, as it must be quite another phenomenon than increased MAO-B activity in brain tissue. Regland et al. (1988) demonstrated that increased activity of platelet MAO is found mainly in SDAT patients, i.e. patients with late onset of dementia. In this subgroup of Alzheimer-demented patients, vitamin B12 deficiency was also shown. The authors considered the high MAO activity in platelets attributable to immature platelets in these not anaemic demented patients. Regland et al. (1989) also showed that patients with pernicious anaemia have very much increased MAO platelet activity, and that the high MAO activity in patients with pernicious anaemia, as well as in SDAT patients, rapidly normalizes when vitamin B12 is instituted.

In an investigation by Thomas et al. (1988), platelet aggregation induced by 5-HT was used as a test of the functional responsiveness of peripheral 5-HT receptors. No significantly different aggregative response was seen in the demented patients.

There are both quantitative and qualitative differences in the changes in the 5-HT system between individuals with normal aging and patients with AD/SDAT. As shown by Carlsson (1986), and as mentioned above, the remaining 5-HT neurons in the normally aged brain can increase metabolism and in this way compensate for the loss of neurons. No such compensatory increase in metabolism has been found in AD/SDAT-afflicted brains. The concentrations of 5-HT, as well as of the end metabolite 5-HIAA, are reduced in demented brains. This may indicate that the remaining 5-HT neurons in the Alzheimer-brain do not function properly.

Investigations of cerebrospinal fluid

In 1969, Gottfries et al. showed that the 5-HIAA concentration in CSF was reduced in patients with SDAT. This finding was later confirmed in probenecid loading tests, showing that the accumulation of 5-HIAA in CSF was reduced in patients with AD/SDAT compared with age-matched groups (Gottfries and Roos, 1973). To date, several investigations of CSF from patients with AD/SDAT have shown reduced concentrations of 5-

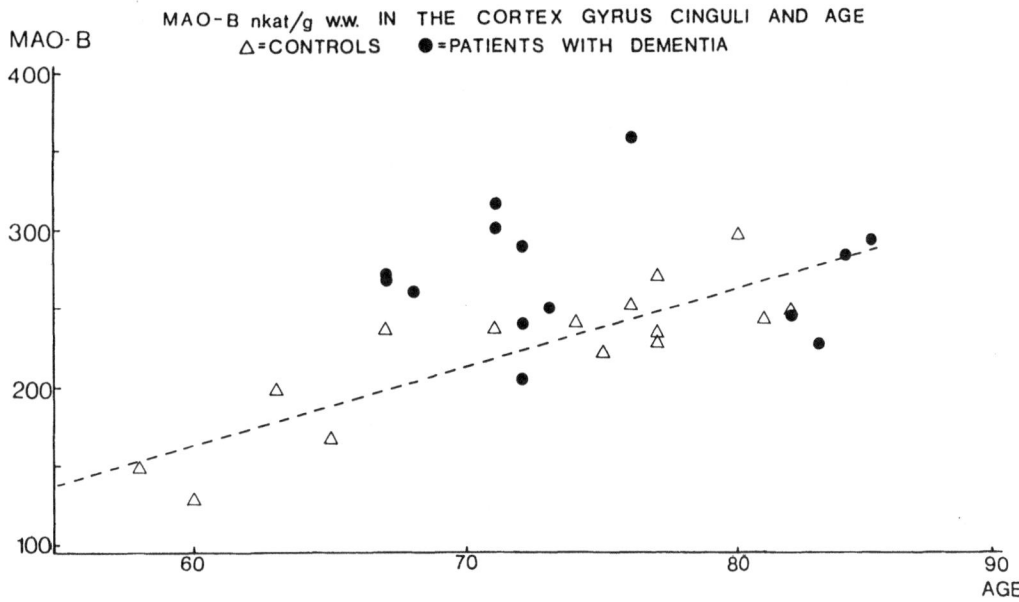

Fig. 4. Increase of monoamine oxidase B in the cortex gyrus cinguli from normally aged individuals and in patients with Alzheimer-type dementia

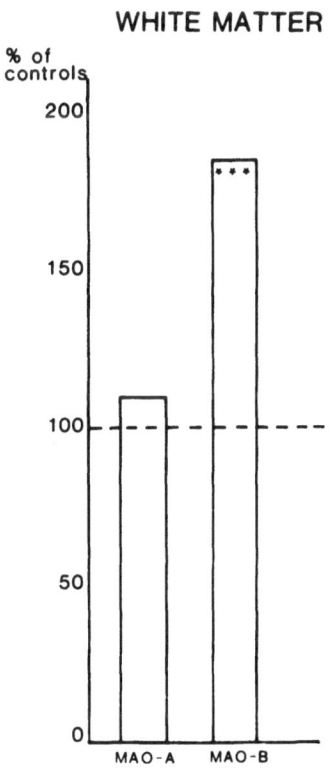

Fig. 5. Increased activity of monoamine oxidase B in white matter from controls and from patients with Alzheimer-type dementia

HIAA, and at our own institute, 123 patients with AD/SDAT were compared with 57 age-matched controls (unpublished data). The mean value of CSF 5-HIAA was 159 ± 63 nmol/l in the controls, and 116 ± 49 nmol/l in the AD/SDAT group. The difference was significant on the 0.0001 level.

Here it should also be mentioned that in an investigation by Forssell et al. (1989) reduced concentrations of tryptophan in CSF were found when AD/SDAT patients were compared with age-matched controls.

Neuroendocrine disturbances and the 5-HT metabolism

As mentioned in the introduction, the 5-HT system is of importance for neuroendocrine function. The system seems to control activity in the hypothalamus. Regarding behavioural disturbances in patients with AD/SDAT, symptoms of sleep disorder, eating disorders and disturbed rhythms around the clock are common. These symptoms indicate a dysfunction of the hypothalamus. Several studies have performed dexamethasone suppression tests (DST) in patients with AD/SDAT. The results vary, but it is obvious that 20–70% of AD/SDAT patients do not respond normally to the test. According to Balldin et al. (1988), it is a combination of stress factors and the dementing process that induces the pathological response to the DST.

In postmortem human brain studies, the hypothalamus was investigated for various biochemical variables (unpublished data). The membrane components phospholipids, cholesterol, gangliosides and sialoglycoproteins were found normal in the hypothalamus. The concentrations of 5-HT and 5-HIAA, however, were significantly reduced in the AD/SDAT group compared with age-matched controls. It was of interest that significantly increased concentrations were found of somatostatin, galanin and arginine vasopressin (AVP) when the neuropeptides were measured in the hypothalamus. The neuropeptide corticotropin-releasing factor (CRF) was also increased, although not significantly. According to Widerlöv et al. (1988), AVP is more potent than CRF as a stimulator of adrenocorticotropic hormone from the pituitary gland. According to these data, there seems to be hyperactivity in the hypothalamus in patients with AD/SDAT, and this may possibly explain the increased activity in the hypothalamic-pituitary-adrenal axis, as marked by the DST.

The correlations were studied between 5-HT and 5-HIAA on one hand and the neuropeptides on the other. Between 5-HT and the neuropeptides there were negative correlations, with r varying between 0.27 and 0.29, but these coefficients were not significant. 5-HIAA, however, was significantly and negatively correlated with somatostatin ($r = -0.42$, $p < 0.01$) and AVP ($r = -0.43$, $p < 0.05$). As it is assumed that 5-HT activity in the hippocampus is of importance for the control of the activity in the hypothalamus, the correlations between the 5-HT variables in the hippocampus and the concentrations of neuropeptides in the hypothalamus were also studied. There were no significant correlations between the hippocampal 5-HT concentration and the concentrations of neuropeptides in the hypothalamus. However, the concentrations of somatostatin and galanin in the hypothalamus were negatively correlated with the concentration of 5-HIAA in the hippocampus ($p < 0.05$). AVP was also negatively correlated with 5-HIAA, but the correlation only bordered on significance. Our data suggest a relationship between the 5-HT input in the hypothalamus and the activity in the HPA axis.

Treatment trials

To further study a possible relationship, patients with AD/SDAT were treated with a selective 5-HT reuptake blocker (citalopram) (Balldin et al., 1988; Nyth et al., 1989). In a subgroup of patients, CSF was drawn before and after four weeks' treatment with citalopram. There was a significant decrease of 5-HIAA in CSF after treatment with citalopram. This was expected, as an increased amount of 5-HT is assumed to accumulate in the synaptic cleft and stimulate the autoreceptors which turn down the synthesis of 5-HT. Still, it is assumed that the total effect of the drug treatment is an up-regulation of the 5-HT system. In the patients treated with citalopram, DST was made before and after four weeks' treatment. The treatment with citalopram significantly decreased the post-dexamethasone cortisol levels. Thus, the pharmacological treatment seems to have inhibited the activity in the HPA axis in patients with AD/SDAT. The clinical study with citalopram was carried out using the double-blind technique, and placebo was used as a reference substance (Nyth et al., 1989). In all, 98 patients were included. The drug treatment caused very few side-effects. No improvement was seen in motor performance or intellectual capacities. However, emotional disturbances, confusion, irritability, restlessness and fear-panic were reduced, and the mood level was raised.

References

Adolfsson R, Gottfries CG, Roos BE, Winblad B (1981) Monoamines in the human brain in normal aging and in senile dementia. In: Marois M (ed) Aging – a challenge to science and social policy, vol 2. Oxford University Press, pp 238–247

Arai H, Kosaka K, Iizuka R (1984) Changes of biogenic amines and their metabolites in postmortem brains from patients with Alzheimer-type dementia. J Neurochem 43:388–393

Balldin J, Gottfries CG, Karlsson I, Lindstedt G, Långström G, Svennerholm L (1988) Relationship between DST and the serotonergic system. Results from treatments with two 5-HT reuptake blockers in dementia disorders. Int J Ger Psychiatry 3:17–26

Balldin J, Gottfries CG, Lindstedt G, Långström G, Svennerholm L (1988) The clonidine growth hormone test in patients with dementia disorders: relation to clinical status and cerebrospinal fluid metabolite levels. Int J Ger Psychiatry 3:115–123

Bucht G, Adolfsson R, Gottfries CG, Roos BE, Winblad B (1981) Distribution of 5-hydroxytryptamine and 5-hydroxyindoleacetic acid in human brain in relation to age, drug influence, agonal status, and circadian variations. J Neural Transm 51:185–203

Carlsson A, Adolfsson R, Aquilonius SM, Gottfries CG, Oreland L, Svennerholm L, Winblad B (1980) Biogenic amines in human brain in normal aging, senile dementia and chronic alcoholism. In: Goldstein M, et al (eds) Ergot compounds and brain function: neuroendocrine and neuropsychiatric aspects. Raven Press, New York, pp 295–304

Carlsson A, Gottfries CG (1986) Neurotransmitter abnormalities in old age dementias. In: Leichner H, Paraschos A (eds) Proceedings of the 5th South-East European Neuropsychiatric Conference, Graz 1983. University Studio Press, Salonica, pp 634–645

Carlsson A (1986) Neurotransmitters in old age and dementia. In: Hafner H, Moschel G, Sartorius N (eds) Mental health in the elderly: a review of the present state of research. Springer, Berlin Heidelberg New York

Forssell LG, Eklöf R, Winblad B (1989) Early stages of late onset Alzheimer's disease II. Derangements in protein metabolism with special reference to tryptophan, tyrosine and cystine. Acta Neurol Scand 79 [Suppl 121]:27–42

Gottfries CG, Gottfries I, Roos BE (1969) Homovanillic acid and 5-hydroxyindoleacetic acid in the cerebrospinal fluid of patients with senile dementia, presenile dementia and Parkinsonism. J Neurochem 16:1341–1345

Gottfries CG, Gottfries I, Johansson B, Olsson R, Persson T, Roos BE, Sjöström R (1971) Acid monoamine metabolites in human cerebrospinal fluid and their relations to age and sex. Neuropharmacology 10:665–672

Gottfries CG, Roos BE (1973) Acid monoamine metabolites in cerebrospinal fluid from patients with presenile dementia (Alzheimer's disease). Acta Psychiatr Scand 49:257–263

Gottfries CG, Roos BE, Winblad B (1976) Monoamine and monoamine metabolites in the human brain post mortem in senile dementia. Aktuelle Gerontologie 6:429–435

Gottfries CG, Adolfsson R, Aquilonius SM, Carlsson A, Eckernäs SE, Nordberg A, Oreland L, Svennerholm L, Wiberg Å, Winblad B (1983) Biochemical changes in dementia disorders of Alzheimer type (AD/SDAT). Neurobiol Aging 4:261–271

Gottfries CG, Bartfai T, Carlsson A, Eckernäs SÅ, Svennerholm L (1986) Multiple biochemical deficits in both gray and white matter of Alzheimer brains. Prog Neuropsychopharmacol Biol Psychiatry 10:405–413

Gottfries CG (1988) Alzheimer's disease – a critical review. Compr Gerontol 2:47–62

Hardy J, Adolfsson R, Alafuzoff I, Bucht G, Marcusson J, Nyberg P, Perdahl E, Wester P, Winblad B (1985) Review. Transmitter deficits in Alzheimer's disease. Critiques: Gottfries CG, Rossor MN, Yates CM. Neurochem Int 7:545–563

Mann DMA, Yates PO (1983) Serotonin nerve cells in Alzheimer's disease. J Neurol Neurosurg Psychiatry 46:96

Marcusson JO, Alafuzoff I, Bäckström IT, Ericson E, Gottfries CG, Winblad B (1987) 5-hydroxytryptamine-sensitive (^{3}H)imipramine binding of protein nature in the human brain. II. Effect of normal aging and dementia disorders. Brain Res 425:137–145

Nagatsu T, Iizuka R (1989) Tyrosine hydroxylase, tryptophan hydroxylase, and the biopterin k-factor in the brains from patients with Alzheimer's disease. J Neural Transm 1:21

Nyth AL, Gottfries CG, Elgen K, Engedal K, Harenko A, Juhela P, Karlsson I, Koskinen T, Nygaard H, Pedersen W, Samuelsson SM, Yli-Kerttula A (1989) The clinical efficacy of citalopram in the treatment of emotional disturbances in dementia disorders. A Nordic multi-centre study

Oreland L, Gottfries CG (1986) Platelet and brain monoamine oxidase in aging and in dementia of Alzheimer's type. Prog Neuropsychopharmacol Biol Psychiatry 10:533–540

Regland B, Gottfries CG, Oreland L, Svennerholm L (1988) Low B12 levels related to high activity of platelet MAO in patients with dementia disorders. A retrospective study. Acta Psychiatr Scand 78:451–457

Regland B, Gottfries CG, Oreland L (1989) Vitamin B12 induced reduction of platelets MAO in patients with dementia and pernicious anaemia (to be published)

Thomas DR, Jones E, Warner N, Harris B, Williams P, Bentley P (1988) Peripheral serotoninergic receptor sensitivity in senile dementia of the Alzheimer type. Biol Psychiatry 23:136–140

Wallin A, Alafuzoff I, Carlsson A, Eckernäs SÅ, Gottfries CG, Karlsson I, Svennerholm L, Winblad B (1989) Neurotransmitter deficits in a non-multi-infarct category of vascular dementia. Acta Neurol Scand 79:397–406

Widerlöv E, Ekman R, Jensen L, Borglund L, Nyman K (1988) Arginine vasopressin, but not corticopropin releasing factor, is a potent stimulator of adrenocorticotropic hormone following electroconvulsive treatment. J Neural Transm 75:101–109

Author's address: Prof. C. G. Gottfries, Department of Psychiatry and Neurochemistry, Gothenburg University, St. Jörgen's Hospital, S-422 03 Hisings Backa, Sweden.

J Neural Transm (1990) [Suppl] 30: 45–55

CSF β-endorphin, HVA and 5-HIAA of dementia of the Alzheimer type and Binswanger's disease in the elderly

S. Lee, T. Chiba, T. Kitahama, R. Kaieda, M. Hagiwara, A. Nagazumi,
and **A. Terashi**

The Second Department of Internal Medicine, Nippon Medical School, Tokyo, Japan

Summary. Cerebrospinal fluid (CSF) concentration of β-endorphin (β-Ep), homovanillic acid (HVA) and 5-hydroxyindoleacetic acid (5-HIAA) was measured in 15 patients with dementia of the Alzheimer type (DAT) and in 16 patients suspected of having Binswanger's disease (BD) by MRI, which sometimes resembles DAT clinically. These were classified into three stages according to severity of dementia, Stage 1 (mild dementia)-Stage 3 (severe dementia). CSF levels of HVA decreased significantly in severe dementia, but the level of 5-HIAA did not correlate with dementia severity in both dementia groups. β-Ep levels did not differ significantly between any stages of DAT, and among controls. β-Ep levels, however, in BD Stage 1 (27.5 ± 5.9 pg/ml) were significantly higher ($p < 0.05$), but level in Stage 3 (6.7 ± 2.0) was significantly lower ($p < 0.001$) than in the controls (19.2 ± 4.5). These results suggest that CSF β-Ep may depend on the cause of dementia rather than severity of dementia, and could possibly distinguish the closely resembling BD from true DAT.

Introduction

Recent advances in radiological equipment have allowed reevaluation of Binswanger's disease (BD) (Binswanger, 1894) which had previously been diagnosable only by autopsy. BD is suspected when a CT scan reveals low density or an MRI reveals hyperintensity in the diffuse periventricular area (Erkinjuntti et al., 1984; Kinkel et al., 1985; Rosenberg et al., 1979; Zeumer et al., 1980). Although slight radiological changes are common in asymptomatic patients (Zimmerman et al., 1986), it has been reported that dementia of the Alzheimer type (DAT) also occasionally shows some periventricular white matter change on CT and MRI (Rezek et al., 1987; George et al., 1986; Steingart et al., 1987). This fact creates confusion between

DAT and BD, as BD sometimes pursues the same course as DAT (Biemond, 1970; Janota, 1981).

In the central nervous system, β-endorphin (β-Ep) is a potential opiate peptide and has several physiological effects: thermoregulation (Tseng et al., 1980), pain perception (Tseng et al., 1980; Clement-Jones et al., 1980), and feeding (Grandison and Giudotti, 1977). It is also suggested that it contributes to learning ability (Izguierdo and Netto, 1985), psychiatric disorders (Almay et al., 1978), cerebrovascular accidents (Furui et al., 1984; Chiba, 1988; Nappi et al., 1986) and others. Several reports describe the relevance of CSF β-Ep to dementia (Kaiya et al., 1983; Facchinetti et al., 1984; Sulkava et al., 1985; Raskind et al., 1986), but their results are inconsistent. CSF levels of HVA: major dopamine metabolite and 5-HIAA: major serotonin metabolite have also been studied in dementia patients, but the results are also inconsistent (Soininen et al., 1981; Bareggi et al., 1982). The purpose of our study is to determine if there is a difference in CSF levels of β-Ep, HVA and 5-HIAA between BD and DAT, and if a correlation can be proven between these levels and the severity of dementia.

Subjects and methods

All subjects or their guardians gave their informed consent for participation in this study. MRI was performed on a 0.15-T Hitachi resistive magnet system. Slice thickness was 10 mm with a 15 mm interslice gap. Spin-echo (SE) images were obtained with a TR of 1200 msec. and TE of 60 msec.

Our study included 16 patients (4 males, 12 females) with BD and 15 patients (4 males, 11 females) with DAT (Table 1). BD was diagnosed by the observation of slowly deteriorating obvious dementia – with or without acute neurological deficits – and MR images which show a bilateral diffuse periventricular hyperintensity (PVH) in the MRI SE sequence (Fig. 1). DAT was diagnosed based on DSM-3-R criteria (American Psychiatric Association, 1987) for primary degenerative dementia and NINCDS-ADRDA diagnostic criteria (McKhan et al., 1984) for probable AD; patients with more than slight PVH were excluded. We performed laboratory tests to exclude inflammatory, metabolic and nutritional causes of dementia. Six age-matched subjects (1 male, 5 females) served as controls (Table 1).

We assessed ischemic score (Hachinski et al., 1975) (Table 1) and Mini-Mental State (MMS) scores (Folstein et al., 1975). Patients were classified into three stages according to dementia severity, as follows (American Psychiatric Association, 1987) (Table 1):

Stage 1. Mild dementia: Definite impairment of memory and of calculating ability are present, but the patient is either fully capable of self-care or needs very little help.

Stage 2. Moderate dementia: Symptoms similar to mild dementia, but impairment of higher functions such as agnosia or apraxia are present and the patient needs assistance.

Stage 3. Severe dementia: Symptoms similar to moderate dementia, but the patient is nearly bedridden, often incontinent and needs much help.

CSF samples were obtained between nine and ten o'clock in the morning after overnight fasting and bed rest in all patients and controls. The level of β-Ep was

Table 1. Comparison of clinical data for patients and controls

	DAT				BD				Controls
	Stage 1	Stage 2	Stage 3	Total	Stage 1	Stage 2	Stage 3	Total	
No. of patients	4	5	6	15	4	6	6	16	6
Age (Mean±SD)	76.3±8.4	78.2±3.0	77.5±10.0	77.4±7.4	74.3±2.6	73.8±3.5	78.2±2.8	75.6±3.5	70.0±9.2
Classification by ischemic score									
DAT	4	5	6	15	2	1	2	5	—
MIX	0	0	0	0	1	2	1	4	—
MID	0	0	0	0	1	3	3	7	—

DAT indicates dementia of the Alzheimer type: *BD* Binswanger's disease; *MID* multi-infarct dementia; *MIX* mixed DAT and MID

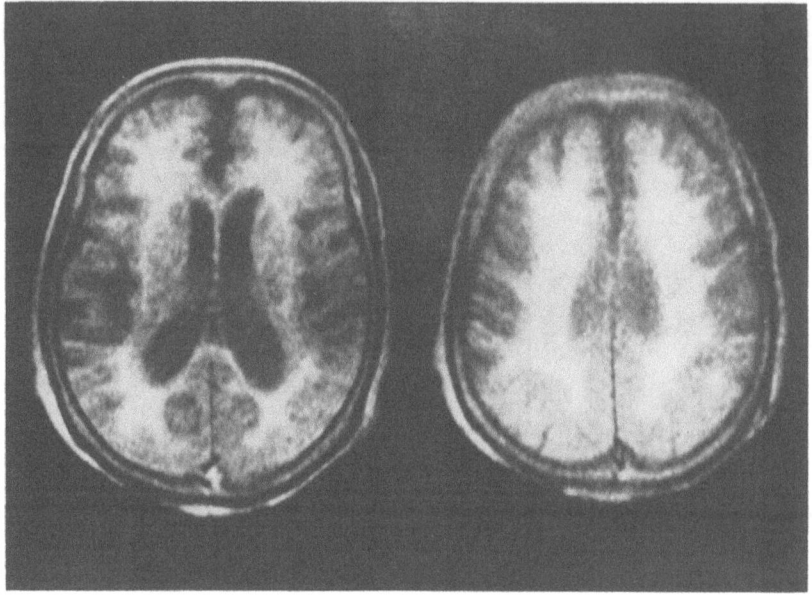

Fig. 1. Transaxial MR images (SE, TR = 1200 msec. TE = 60 msec.) obtained at 0.15 T. Diffuse periventricular hyperintensity (PVH) is recognized in 78 year old woman with severe dementia

measured by the radioimmunoassay method (RIA) as previously described (Furui et al., 1984; Yasunari et al., 1985). The antibody for RIA was raised in rabbits by immunization with synthetic β-Ep in complete Freund's adjuvant. The assay shows only a 3.3% cross-reaction with β-lipotropin and no cross reaction with α-endorphin, methionine- and leucine-enkephalin. HVA and 5-HIAA were measured by high performance liquid chromatography (HPLC) method.

The statistical significance of the differences between patients and controls was evaluated by Student's t-test. A p value above 0.05 was considered not significant.

Results

The mean ages of patient groups and controls did not differ significantly (Table 1). The CSF level of HVA in DAT stage 3 (32.3 ± 15.7 ng/ml, Mean $\pm$ SD) was significantly lower ($p < 0.05$) than the controls (73.0 ± 12.0) (Fig. 2), but there was no correlation with MMS scores (Fig. 3). Likewise the level in BD stage 3 (37.5 ± 20.8) was significantly lower ($p < 0.05$) than the controls (Fig. 4), and there was a weak correlation between these levels and MMS scores (Fig. 5). The CSF level of 5-HIAA in DAT Stage 2 showed a significantly lower level (16.6 ± 3.8 ng/ml, $p < 0.05$) than the controls (27.7 ± 8.1) (Fig. 6), but there was no correlation between such levels and MMS scores. CSF 5-HIAA levels in BD were not significantly different than the controls (Fig. 7), and there was no correlation between such levels and MMS scores either. The CSF levels of β-Ep did not differ

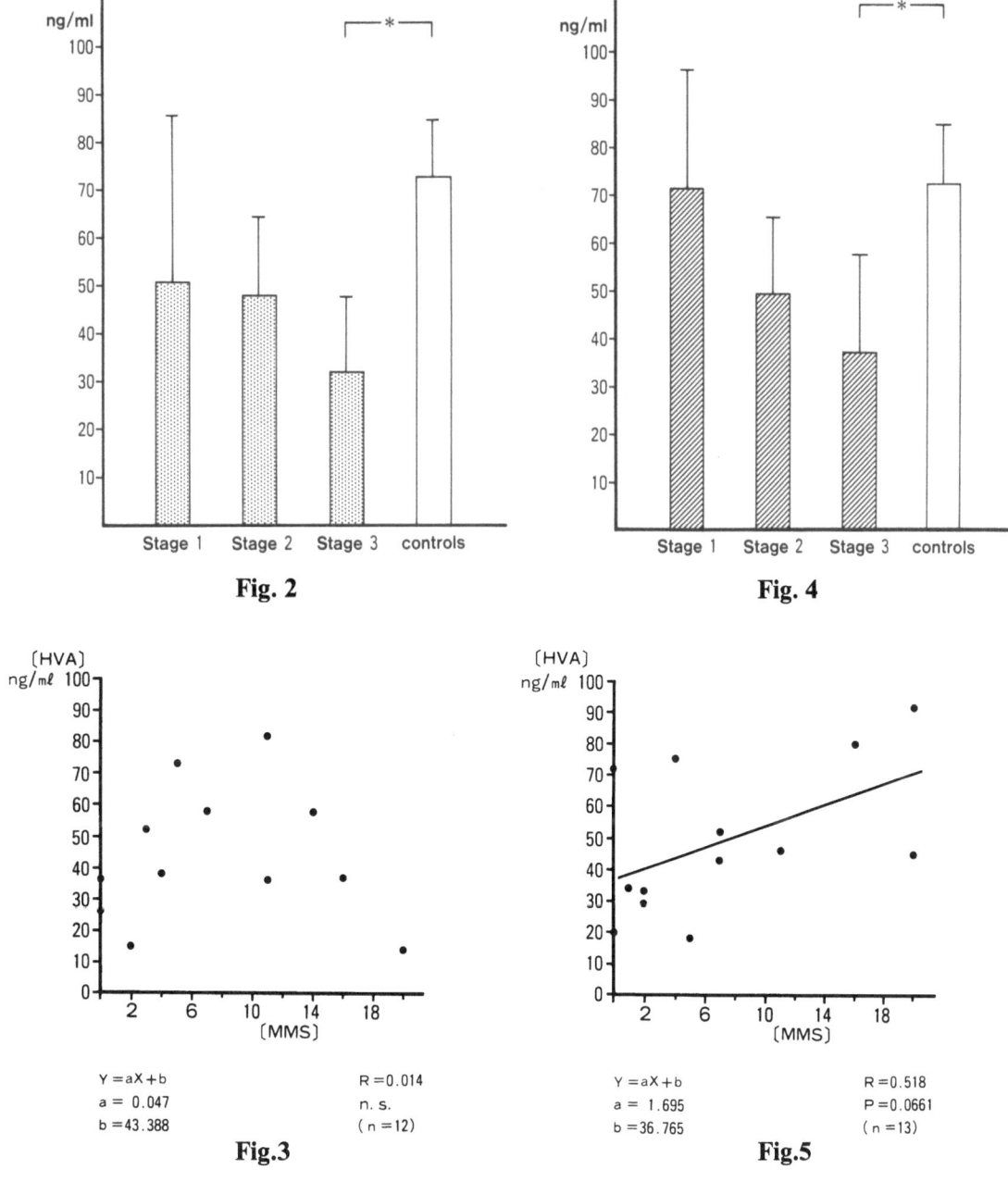

Fig. 2. Mean (±SD) CSF levels of HVA between various dementia stages in DAT and controls (* p < 0.05)

Fig. 3. Correlation between the CSF levels of HVA and MMS scores in DAT

Fig. 4. Mean (±SD) CSF levels of HVA between various dementia stages in BD and controls (* p < 0.05)

Fig. 5. Correlation between the CSF levels of HVA and MMS scores in BD

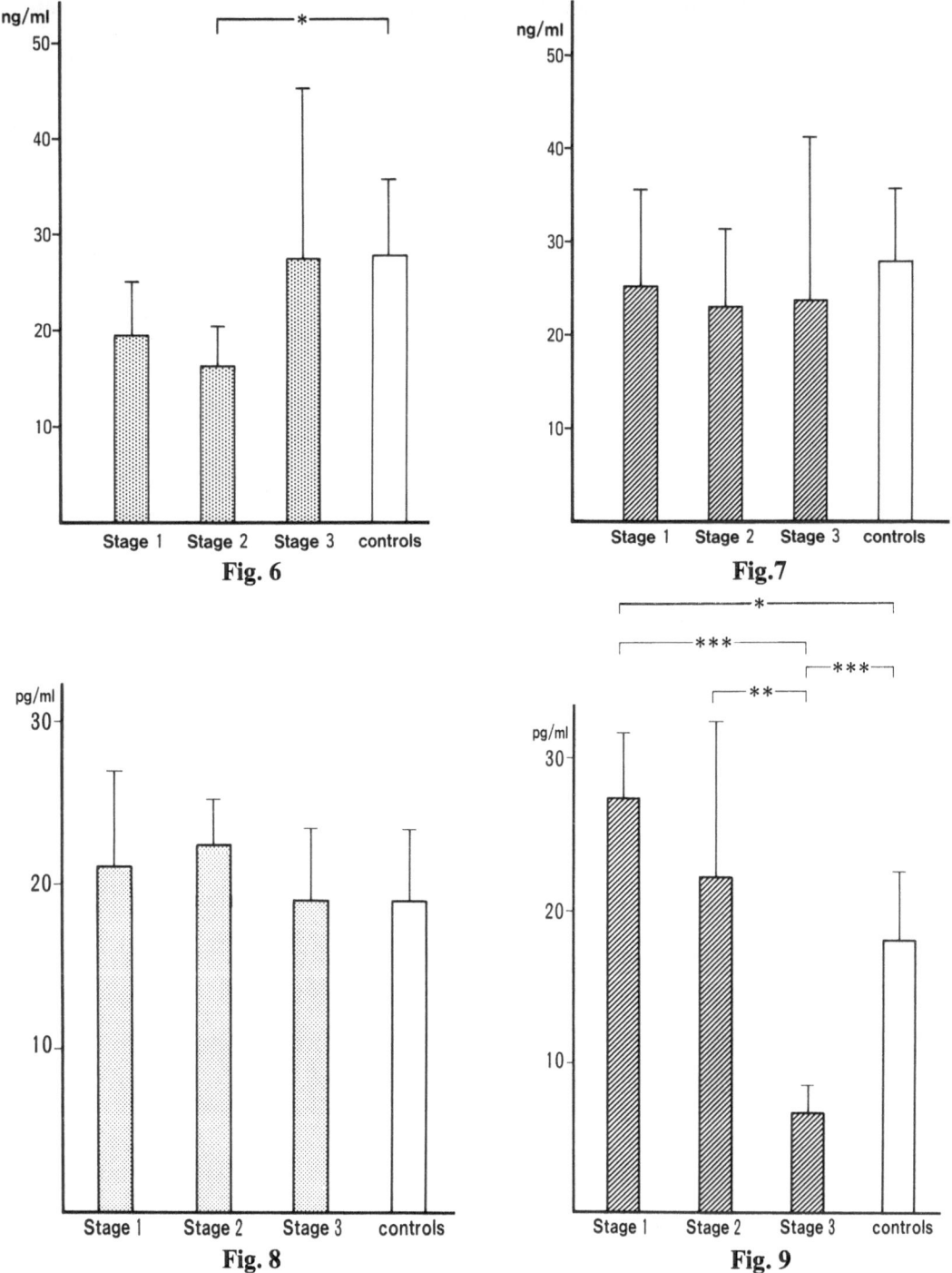

Fig. 6. Mean (± SD) CSF levels of 5-HIAA between various dementia stages in DAT and controls (* p < 0.05)

Fig. 7. Mean (± SD) CSF levels of 5-HIAA between various dementia stages in BD and controls

Fig. 8. Mean (± SD) CSF levels of β-Ep between various dementia stages in DAT and controls

Fig. 9. Mean (± SD) CSF levels of β-Ep between various dementia stages in BD and controls (* p < 0.05; ** p < 0.01; *** p < 0.001)

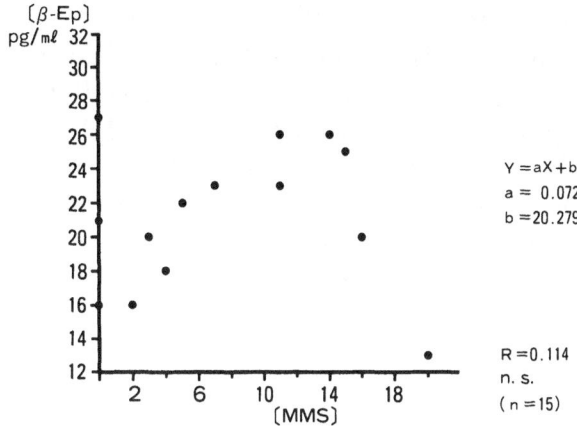

Fig. 10. Correlation between the CSF levels of β-Ep and MMS scores in DAT

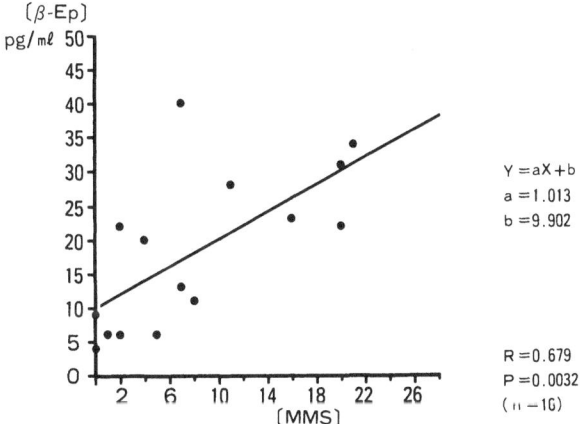

Fig. 11. Correlation between the CSF levels of β-Ep and MMS scores in BD

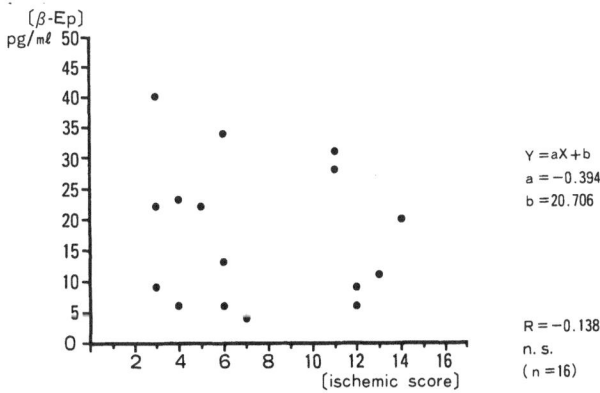

Fig. 12. Correlation between the CSF levels of β-Ep and ischemic scores in BD

significantly between DAT stages (Stage 1: 21.0 ± 5.9 pg/ml, Stage 2: 22.4 ± 2.9, Stage 3: 19.3 ± 4.4) and controls (19.2 ± 4.5) (Fig. 8). The CSF level of β-Ep was, however, significantly higher ($p < 0.05$) in BD Stage 1 (27.5 ± 5.9) and significantly lower ($p < 0.001$) in BD Stage 3 (6.7 ± 2.0) than in the controls. The level in BD Stage 2 (22.3 ± 10.6) did not show a significant difference from the controls (Fig. 9). Between CSF β-Ep levels and MMS scores there was no correlation in DAT (Fig. 10), but there was a strong correlation in BD (Fig. 11). Between ischemic score and β-Ep levels in BD, no correlation was recognized (Fig. 12).

Discussion

The results of the present study suggest that in mild dementia caused by BD, CSF β-Ep levels are increased, whereas in severe dementia these are decreased; in DAT, however, there seems to be no correlation between CSF β-Ep levels and dementia severity. Our results could not confirm previous reports which showed low CSF β-Ep levels in DAT (Kaiya et al., 1983; Facchinetti et al., 1984). One of the reasons might be the differences of age between the subjects and controls in the previous reports, but this correlation – between aging and CSF β-Ep levels – is still controversial (Raskind et al., 1986; Facchinetti et al., 1983). Another reason might be a difference in patient selection. There is also some difficulty in DAT diagnosis: several reports describe the discrepancy between clinical and pathological diagnosis of DAT (Todorov et al., 1975; Wade et al., 1987). If patients are selected more exclusively, it causes the elevation of specificity with a loss of sensitivity. In our study, DAT patients were rather highly selected by diagnostic criteria and laboratory tests including MRI. Several reports show that patients with Alzheimer type dementia also show periventricular white matter change on MRI and CT (Rezek et al., 1987; George et al., 1986; Steingart et al., 1987). We excluded, however, patients who showed excessive PVH because of possible vascular disorders that could cause misdiagnosis between DAT and BD. This relatively low sensitivity, however, might be related to the CSF β-Ep levels.

BD is reported to show a clinically slow progressive intellectual deterioration with focal neurological events (Jellinger and Neumayer, 1964; Olszewski, 1965; Caplan and Schoene, 1978), but sometimes its course resembles DAT (Biemond, 1970; Janota, 1981). It is stressed that characteristic pathological findings are diffuse demyelination in white matter with multiple infarcts and relative sparing of cortex and U-fibers. The etiology is generally thought to be hypoxia or circulatory insufficiency due to arteriolar sclerosis of long penetrating arteries (Rosenberg et al., 1979; Zeumar et al., 1980; Jellinger and Neumayer, 1964; Olszewski, 1965; Caplan and Schoene, 1978). β-Ep is mainly located in the hypothalamus, but it is also widely distributed in the cerebrum by nerve fibers. This fact gives rise to a hypothesis that β-Ep elevation in BD Stage 1: could be stimulated by hypoxia and acidosis just as in the umbilical cord plasma of term human fetuses (Wardlaw et al., 1979). On the other hand a sodium disorder in the white matter due to demyelination (Brismar, 1983; Bourre et al., 1982), may contribute to the release of brain β-Ep because sodium ions inhibit agonist interaction at opioid receptors (Frances et al., 1985). The reason for the decrease of the CSF β-Ep levels in patients in Stage 3 of BD could be that demyelination causes a loss of connection between the cerebral hemisphere and the hypothalamus.

CSF levels of HVA and 5-HIAA in Alzheimer's disease are also controversial (Soininen et al., 1981; Bareggi et al., 1982). In our study, the HVA level decreased in the poor clinical stage in both dementia groups. Such levels, however, did not correlate well with MMS scores in DAT, and in BD the correlation was weak. One of the functions of dopamine is probably a regulation of motor functions. The patients' bedridden state in progressed dementia stage may relate to diminution of HVA, the main metabolite of dopamine. CSF 5-HIAA may derive from the cerebrum and also from the spinal cord. The rate from the spinal cord is reported more than 30% (Bulat et al., 1974; Bulat, 1977), and this may have caused no significant correlation between the levels and severity of dementia.

In conclusion, CSF β-Ep level was different according to dementia severity in BD but not in DAT. The CSF β-Ep level may relate with the cause of dementia rather than severity and could possibly distinguish the closely resembling BD from true DAT.

Acknowledgement

The authors are especially grateful to the late Kouichi Iyoda, Ph. D., for invaluable suggestions.

References

Almay BG, Johansson F, von Knorring L, Terenius L, Wahlström A (1978) Endorphins in chronic pain. 1. Differences in CSF endorphin levels between organic and psychogenic pain syndromes. Pain 5:153–162

American Psychiatric Association (1987) Diagnostic and statistical manual of mental disorders, 3rd edn (revised). Washington, DC

Bareggi SR, Franceschi M, Bonini L, Zecca L, Smirne S (1982) Decreased CSF concentrations of homovanillic acid and γ-aminobutyric acid in Alzheimer's disease: age or disease-related modifications? Arch Neurol 39:709–712

Biemond A (1970) On Binswanger's subcortical arteriosclerotic encephalopathy and possibility of its clinical recognition. Psychiat Neurol Neurochir 73:413–417

Binswanger O (1894) Die Abgrenzung der allgemeinen progressiven Paralyse. Berl Klin Wochenschr 31:1103–1105, 1137–1139, 1180–1186

Bourre JM, Chanez C, Dumont O, Flexor MA (1982) Alteration of 5′-nucleotidase and Na⁺, K⁺-ATPase in central and peripheral nervous tissue from dysmyelinating

54   S. Lee et al.

mutants (jimpy, quaking, Trembler, shiverer, and mld). Comparison with CNPase
  in the developing sciatic nerve from Trembler. J Neurochem 38:643–649
Brismar T (1983) Neuropathy-functional abnormalities in the BB rat. Metabolism 32
  [Suppl 1]:112–117
Bulat M (1977) On the cerebral origin of 5-hydroxyindoleacetic acid in the lumbar
  cerebrospinal fluid. Brain Res 122:388–391
Bulat M, Lackovic Z, Jakupcevic M (1974) 5-Hydroxyindoleacetic acid in the lumbar
  fluid: a specific indicator of spinal cord injury. Science 185:527–528
Caplan LR, Schoene WC (1978) Clinical features of subcortical arteriosclerotic en-
  cephalopathy (Binswanger's disease). Neurology 28:1206–1215
Chiba T (1988) Studies of β-endorphin and methionin-enkephalin in cerebral vascular
  disease. Its evaluation and clinical significance of periodic changes. Nippon Ika
  Daigaku Zasshi 55:46–53
Clement-Jones V, McLoughlin L, Tomlin S, Besser GM, Rees LH, Wen HL (1980)
  Increased β-endorphin but not met-enkephalin levels in human cerebrospinal fluid
  after acupuncture for recurrent pain. Lancet ii:946–949
Erkinjuntti T, Sipponen JT, Iivanainen M, Ketonen L, Sulkava R, Sepponen RE
  (1984) Cerebral NMR and CT imaging in dementia. J Comput Assist Tomogr
  8:614–618
Facchinetti F, Nappi G, Petraglia F, Martignoni E, Sinforiani E, Genazzani AR
  (1984) Central ACTH deficit in degenerative and vascular dementia. Life Sci
  35:1691–1697
Facchinetti F, Petraglia G, Nappi G, Martignoni E, Antoni G, Parrini D, Genazzani
  AR (1983) Different patterns of central and peripheral βEP, βLPH and ACTH
  throughout life. Peptides 4:469–474
Folstein MF, Folstein SE, McHugh PR (1975) "Mini-Mental State" A practical
  method for grading the cognitive state of patients for the clinician. J Psychiatr Res
  12:189–198
Frances B, Moisand C, Meunier J-C (1985) Na$^+$ ions and Gpp(NH)p selectively
  inhibit agonist interactions at μ- and k-opioid receptor sites in rabbit and Guinea-
  pig cerebellum membranes. Eur J Pharmacol 117:223–232
Furui T, Satoh K, Asano Y, Shimosawa S, Hasuo M, Yaksh TL (1984) Increase of
  β-endorphin levels in cerebrospinal fluid but not in plasma in patients with cerebral
  infarction. J Neurosurg 61:748–751
George AE, de Leon MJ, Kalnin A, Rosner L, Goodgold A, Chase N (1986) Leukoen-
  cephalopathy in normal and pathologic aging. 2. MRI of brain lucencies. AJNR
  7:567–570
Grandison L, Guidotti A (1977) Stimulation of food intake by muscimol and beta
  endorphin. Neuropharmacology 16:533–536
Hachinski VC, Iliff LD, Zilhka M, Du Boulay GH, McAllister VL, Marshall J, Ross
  Russell RW, Symon L (1975) Cerebral blood flow in dementia. Arch Neurol
  32:632–637
Izquierdo I, Netto CA (1985) The brain β-endorphin system and behavior: the mod-
  ulation of consecutively and simultaneously processed memories. Behav Neural
  Biol 44:249–265
Jellinger K, Neumayer E (1964) Progressive subcorticale vasculäre Encephalopathie
  Binswanger. Eine klinisch-neuropathologische Studie. Arch Psychiatr Nervenkr
  205:523–554
Janota I (1981) Dementia, deep white matter damage and hypertension: 'Bins-
  wanger's disease'. Psychol Med 11:39–48
Kaiya H, Tanaka T, Takeuchi K, Morita K, Adachi S, Shirakawa H, Ueki H, Namba
  M (1983) Decreased level of β-endorphin-like immunoreactivity in cerebrospinal
  fluid of patients with senile dementia of Alzheimer type. Life Sci 33:1039–1043

Kinkel WR, Jacobs L, Polachini I, Bates V, Heffner RR Jr (1985) Subcortical arteriosclerotic encephalopathy (Binswanger's disease). Computed tomographic, nuclear magnetic resonance, and clinical correlations. Arch Neurol 42:951–959

McKhan G, Drachman D, Folstein M, Katzman R, Price D, Stadlan E (1984) Clinical diagnosis of Alzheimer's disease: report of the NINCDS-ADRDA Work Group under the auspices of Department of Health and Human Services Task Force on Alzheimer's disease. Neurology 34:939–944

Nappi G, Facchinetti F, Bono G, Petraglia F, Sinforiani E, Genazzani AR (1986) CSF and plasma levels of pro-opiomelanocortin-related peptides in reversible ischemic attacks and strokes. J Neurol Neurosurg Psychiatry 49:17–21

Olszewski J (1965) Subcortical arteriosclerotic encephalopathy. Review of the literature on the so-called Binswanger's disease and presentation of two cases. World Neurol 3:359–374

Raskind MA, Peskind ER, Lampe TH, Risse SC, Taborsky GJ Jr, Dorsa D (1986) Cerebrospinal fluid vasopressin, oxytocin, somatostatin, and β-endorphin in Alzheimer's disease. Arch Gen Psychiatry 43:382–388

Rezek DL, Morris JC, Fulling KH, Gado MH (1987) Periventricular white matter lucencies in senile dementia of Alzheimer type and in normal aging. Neurology 37:1365–1368

Rosenberg GA, Kornfeld M, Stovring J, Bicknell JM (1979) Subcortical arterisclerotic encephalopathy (Binswanger). Computerized tomography. Neurology 29:1102–1106

Soininen H, MacDonald E, Rekonnen M, Riekkinen PJ (1981) Homovanillic acid and 5-hydroxyindoleacetic acid levels in cerebrospinal fluid of patients with senile dementia of Alzheimer type. Acta Neurol Scand 64:101–107

Steingart A, Hacinski VC, Lau C, Fox AJ, Fox H, Lee D, Inzitari D, Merskey H (1987) Cognitive and neurologic findings in demented patients with diffuse white matter lucencies on computed tomographic scan (Leuko-Araiosis). Arch Neurol 44:36–39

Sulkava R, Erkinjuntti T, Laatikainen T (1985) CSF β-endorphin and β-lipotropin in Alzheimer's disease and multi-infarct dementia. Neurology 35:1057–1058

Todorov A, Go RC, Constantinidis J, Elston R (1975) Specificity of the clinical diagnosis of dementia. J Neurol Sci 26:81–98

Tseng LF, Wei ET, Loh HH, Li CH (1980) β-Endorphin: central sites of analgesia, catalepsy and body temperature changes in rats. J Pharm Exp 214:328–332

Wade JP, Mirsen TR, Hachinski VC, Fisman M, Lau C, Merskey H (1987) The clinical diagnosis of Alzheimer's disease. Arch Neurol 44:24–29

Wardlaw SL, Stark RI, Baxi L, Frantz AG (1979) Plasma β-endorphin and β-lipotropin in the human fetus at delivery: correlation with arterial pH and pO_2. J Clin Endocrinol Metab 49:888–891

Yasunari K, Kohno M, Kanayama Y, Kono Y, Amatsu K, Takeda T, Sato K, Kotsugi N (1985) Changes of plasma levels of β-endorphin-like immunoreactivity after acute clonidine administration in patients with essential hypertension. Horm Metabol Res 17:324–325

Zeumer H, Schonsky B, Sturm KW (1980) Predominant white matter involvement in subcortical arteriosclerotic encephalopathy (Binswanger disease). J Comput Assist Tomogr 4:14–19

Zimmerman RD, Flemming CA, Lee BCP, Saint-Louis LA, Deck MDF (1986) Periventricular hyperintensity as seen by magnetic resonance. Prevalence and significance. AJR 146:443–450

Authors' address: S. Lee, M.D., The Second Department of Internal Medicine, Nippon Medical School, 3-5-5 Iidabashi Chiyoda-ku, Tokyo 102, Japan.

J Neural Transm (1990) [Suppl] 30: 57–67

Neuropeptide levels in Alzheimer's disease and dementia with frontotemporal degeneration

L. Minthon[1], **L. Edvinsson**[2], **R. Ekman**[3], and **L. Gustafson**[1]

[1] Departments of Psychogeriatrics, [2] Internal Medicine,
and [3] Psychiatry and Neurochemistry, University of Lund, Lund, Sweden

Summary. The CSF levels of somatostatin-LI (SLI), neuropeptide Y (NPY-LI) and Delta Sleep Inducing Peptide (DSIP-LI) have been measured in patients with dementia of Alzheimer type (DAT) and dementia with frontotemporal degeneration of non-Alzheimer type (FTD). The distribution pattern of cortical degeneration differs between these two types of dementia. DAT shows degeneration of mainly temporo-parietal and temporo-limbic structures, whereas FTD discloses its main degeneration in the frontotemporal regions (Brun, 1987). The somatostatin-LI was significantly reduced both in DAT and FTD. NPY-LI showed a significant reduction in DAT but not in FTD. A tendency to a reduction with duration of the disease was observed in DAT whereas the contrary was noted in FTD. The DSIP-LI levels were reduced in DAT and slightly increased in FTD. The study provides an evidence of neurochemical differences between the two primary degenerative dementias.

Introduction

Dementia of Alzheimer type (DAT) is the most common primary degenerative dementia, responsible for about 50% of all severe cases of dementia (Gustafson and Brun, 1989). The pattern of cortical involvement found in most DAT cases is an accentuation of the degeneration in the temporo-parietal association cortex and in temporal-limbic structures (Brun and Gustafson, 1976). This degenerative pattern is almost contrary to that found in dementia with fronto-temporal cortical degeneration of non-Alzheimer type (FTD including Pick's disease). FTD is the second most common type of primary degenerative dementia in our catchment area, and found in 10% of a consecutive series of post mortem verified cases.

The disease process in FTD affects mainly frontal and anterior temporal regions and to a less extent postcentral cortical areas. Also the histopathology of DAT and FTD differs in many aspects (Brun, 1987). Clinical differentiation between DAT and FTD is difficult especially at an early stage, but becomes more apparent in the progress of the diseases (Gustafson, 1987). The available information on peptides such as somatostatin, neuropeptide Y (NPY) and Delta Sleep Inducing Peptide (DSIP) is scarce in DAT and FTD. A reduction in DAT somastatin-like immunoreactivity (SLI) and NPY-LI has been reported. This has been shown in the brain tissues or in the cerebrospinal fluid (CSF) (Davies et al., 1980; Rossor et al., 1980; Beal et al., 1987; Nakamura et al., 1986). The present study was carried out in order to examine differences in CSF levels of SLI, NPY-LI and DSIP-LI in the two types of primary degenerative dementia, DAT and FTD. The patients were carefully examined with all available clinical techniques to ascertain the diagnoses, and the results were compared with CSF findings in a group of healthy subjects.

Materials and methods

Consecutive cases with a clinical picture of DAT or FTD were selected among patients referred to the Psychogeriatric Department for diagnostic examination and treatment. All patients went through psychiatric, psychometric, physical and laboratory examinations, regional cerebral blood flow (rCBF) measurement, EEG and in most cases a brain computerized tomography (CT). RCBF was measured with the ^{133}Xe inhalation technique measuring 32 or 254 regions of both hemispheres (Risberg, 1980; Risberg, 1987).

The clinical diagnoses of DAT and FTD were based on the neuropsychiatric examinations including the rCBF findings. The diagnosis was supported by scores in three diagnostic rating scales: ischemic score (IS) (Hachinski et al., 1975); a rating scale for diagnosis of DAT (Gustafson and Nilsson, 1982), and a rating scale for diagnosis of FTD (Gustafson and Nilsson, 1982). The clinical diagnoses based on these rating scales, in combination, have been validated against post mortem diagnoses and rCBF findings (Brun and Gustafson, 1988; Risberg et al., 1983; Gustafson and Brun, 1989). The accuracy of differential diagnosis by rCBF measurements is also high as shown by comparison with post mortem diagnoses (Gustafson et al., 1984). Patients with vascular dementia, organic brain syndromes with known etiology, chronic psychosis, addiction, or severe somatic disease were excluded from the study.

Eight patients died during the follow-up period. The clinical diagnosis was confirmed in seven (4 FTD and 3 DAT) cases in which a post mortem investigation was performed.

The reference group consisted of 11 healthy volunteers (6 females, 5 males) with a mean age of 40 years (range 23–68 years). None of the subjects had any signs of neurological disorder or dementia according to the clinical judgement performed, nor had they received any continual medication.

This study was approved by the Ethic's committee of the University of Lund.

CSF sampling

The levels of neuropeptides were analysed in CSF that was removed directly into vials containing EDTA. The first three millilitres were discarded and the following ten saved. The samples were transported on ice to the laboratory for centrifugation at $+4°C$ for 10 min, and then stored frozen at $-70°C$ for later analysis.

Radioimmunoassay

1. Neuropeptide Y

For the radioimmunoassay (RIA) of NPY a rabbit antiserum was raised against synthetic porcine NPY (a kind gift from Dr. P.C. Emson, Cambridge, England) conjugated to bovine serum albumin with carbodiimide. ^{125}I-NPY, used as a tracer, was purified by high performance liquid chromatograpy (HPLC). The antiserum cross-reacted with peptide YY (PYY) to an extent of 33% but not with C-terminal fragments of NPY and PYY (NPY 13–36 and PYY 13–36) nor with bovine pancreatic polypeptide, gastrin inhibiting peptide, peptide histidyl isoleucine, vasoactive intestinal peptide or secretin (Widerlöv et al., 1988).

Two hundred μl of antiserum (diluted 1:40 000) were incubated first with 100 μl of standard (synthetic NPY; Peninsula, Belmont, CA, USA) or CSF for 24 h at $4°C$ and with 200 μl (about 2500 cpm) of the HPLC purified tracer for another 24 h. Bound and free ^{125}I-NPY were separated using dextran-coated charcoal. Each CSF sample was assayed in serial dilution and corrected for nonspecific binding. The detection limit was 11.7 pmol/l. Intraassay variation was 6.5% and interassay variation 7.0%.

Because of cross-reactivity with PYY, each sample was also assayed for PYY-LI and 33% of the values obtained were subtracted from the corresponding values for NPY-LI. A rabbit antiserum (K-8413), used in a previously described PYY-RIA (Ekman et al., 1986), recognized the N-terminal portion of the PYY-molecule and does not cross-react with synthetic porcine NPY. The minimum detectable amount of authentic porcine PYY was 9 pmol/l and the intra- and interassay variations were 4.9 and 5.8%, respectively, in the range 20–140 pmol/l.

2. Somatostatin

Immunoreactive somatostatin was quantitated using a radioimmunoassy published previously (Wallengren et al., 1987). The somatostatin antiserum (K18, Milab, Malmö, Sweden) was used in a final dilution of 1:25 000. It does not cross-react with any other known neuropeptide besides cyclic somatostatin (100%), linear somatostatin (50%), [tyr-^{1}]-somatostatin (100%), and [tyr-^{1}]-somatostatin (38%). Two hundred microliters of antiserum were incubated first with 100 μl of standard or extract for 24 h at $4°C$ and then with 200 μl (4000 cpm) of ^{125}I-[tyr]-somatostatin for another 24 h at $4°C$. Bound and free ^{125}I-[tyr-^{1}]-somatostatin were separated using dextran-coated charcoal (0.5% activated charcoal, 0.1% Dextran T-70 in 0.05 M phosphate buffer, pH 7.5 containing 0.25% human serum albumin). The radioactivity of the supernatant was counted in an LKB 1260 gamma counter. The detection limit was 5 pmol/l; intraassay variation was 6% and interassay variation was 12%. All samples were assayed in serial dilutions (duplicate samples).

3. Delta sleep inducing peptide

A highly sensitive radioimmunoassay for DSIP has been developed at the laboratory (Ekman et al., 1983). A p-hydroxyphenylpropionic acid conjugate of DSIP was used for radioiodination. Using reversed-phase high performance liquid chromatography the labelled DSIP derivative was isolated in a high yield and with a high specific activity. The assay allows measurement of DSIP-like material with a minimum detectable concentration of 0.1 ng/ml standard DSIP (10 pg/tube).

Results and comments

Neuropeptide Y

Age characteristics and diagnostic score in patients with DAT and FTD are listed in Table 1. The NPY-LI levels in CFS were anlysed as described above (Fig. 1). The mean NPY-LI level in DAT (109 ± 21 pmol/l) was significantly lower than in normals (139 ± 12 pmol/l, $p < 0.05$) and especially when compared to FTD (154 ± 36 pmol/l, $p < 0.001$). NPY-LI in FTD was mainly within normal range. There were no significant correlations between NPY-LI and the duration of dementia in DAT and FTD. However in DAT the NPY-LI tended to decrease with longer duration of the disease (Fig. 2). There was no significant correlation between NPY-LI and the age at onset of dementia in DAT, but there were different trends in DAT and FTD. In DAT NPY-LI showed a decrease and in FTD an increase with age at onset and also with the duration of the disease (Fig. 2). The differences between DAT and FTD might indicate that the changes in cerebral cortical NPY neurons during the course of primary degenerative dementias, might be related to the type of the disease and its involvement of certain brain areas. Thus, the Alzheimer disease mainly involves the temporo-limbic and the temporo-parietal cortex, and the posterior part of the cingulate gyrus (Brun and Englund, 1981). These may be the areas where the NPY neurons show degeneration. In contrast the degeneration

Table 1. Age characteristics and diagnostic score in patients with DAT and FTD, and in normals

	n	♂	♀	Diagnostic score			Age at onset	Age at invest.	Dura-tion
				DAT	FTD	IS			
DAT	17	9	8	7.9 ± 1.4	1.9 ± 0.7	1.2 ± 0.4	58.3 ± 6.3	62.6 ± 6.2	4.3 ± 1.9
FTD	11	4	7	3.1 ± 1.1	6.5 ± 1.6	1.6 ± 1.4	52.7 ± 10.5	56.5 ± 11.7	3.8 ± 2.5
Normals	11	5	6	–	–	–	–	40 ± 14	–

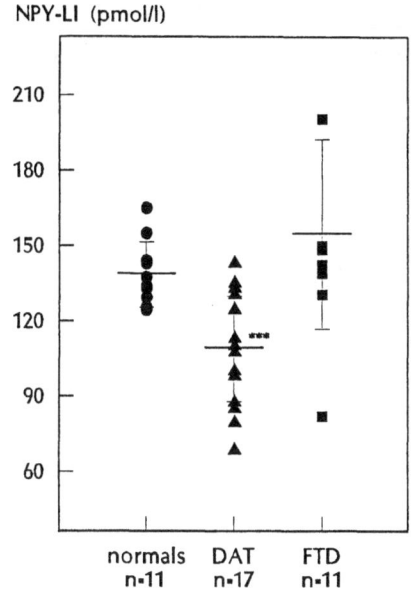

Fig. 1. Mean CSF levels of NPY-LI in patients with dementia of Alzheimer type (DAT) and frontotemporal degeneration (FTD), and in normals

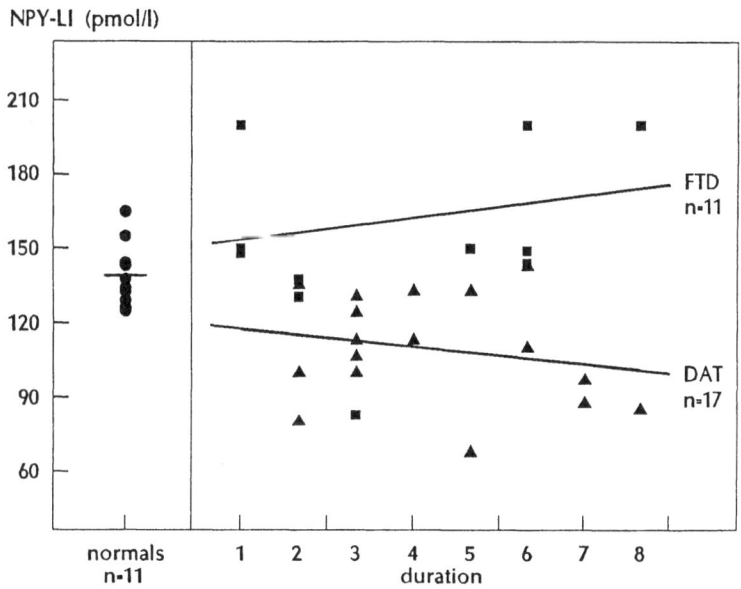

Fig. 2. The relationship between CSF NPY-LI and duration of dementia of Alzheimer type (DAT) and of frontotemporal degeneration (FTD)

in FTD has its maximum impact in the frontal and the anterior temporal cortex (Brun, 1987). These areas may seem to contain abundant NPY neurons. We presume instead an upregulation of the NPY neuron, reflected on the levels in CSF.

Somatostatin

Age characteristics and diagnostic score in patients with dementia of Alzheimer type (DAT), fronto-temporal degeneration (FTD) and normals are given in Table 2. There was a clearcut difference between the scoring profiles in DAT and FTD, concerning DAT and FTD scores, but not in IS. The mean level of SLI was significantly reduced from 54 ± 24 pmol/l in normals to 29 ± 13 pmol/l in DAT and 27 ± 12 pmol/l in FTD ($p < 0.001$). There was no significant difference between the two patient groups (Fig. 3), and there were no significant correlations between the duration of the diseases and the somatostatin levels in the two degenerative dementias (Fig. 4). This is the first demonstration that the SLI is reduced in FTD as well as in DAT.

Delta sleep inducing peptide

Age characteristics and diagnostic score in patients with dementia of Alzheimer type (DAT) and frontotemporal degeneration (FTD) examined

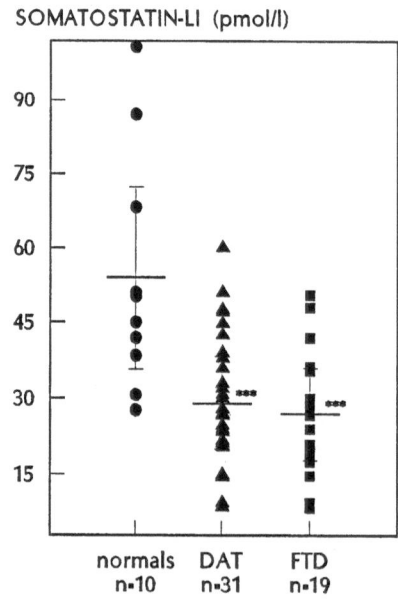

Fig. 3. Mean CSF levels of somatostatin-LI in patients with dementia of Alzheimer type (DAT) and frontotemporal degeneration (FTD), and in normals

Table 2. Age characteristics and diagnostic score in patients with DAT and FTD, and in normals

	n	♂	♀	Diagnostic score			Age at onset	Age at invest.	Duration
				DAT	FTD	IS			
DAT	31	15	16	7.6±1.4	2.0±0.7	1.5±0.7	60± 7	64± 7	4±2
FTD	19	8	11	2.9±1.0	6.6±1.8	1.7±1.3	54±10	58±10	4±2.6
Normals	10	4	6	–	–	–	–	37±10	–

Table 3. Age characteristics and diagnostic score in patients with DAT and FTD, and in normals

	n	♂	♀	Diagnostic score			Age at onset	Age at invest.	Duration
				DAT	FTD	IS			
DAT	30	15	15	7.6±1.5	2.0±7	1.5±7	60±7	64± 7	4±2.0
FTD	21	8	13	2.9±1.0	1.7±1.3	1.7±1.3	54±9	58±10	4±2.6
Normals	11	5	6	–	–	–	–	40±14	–

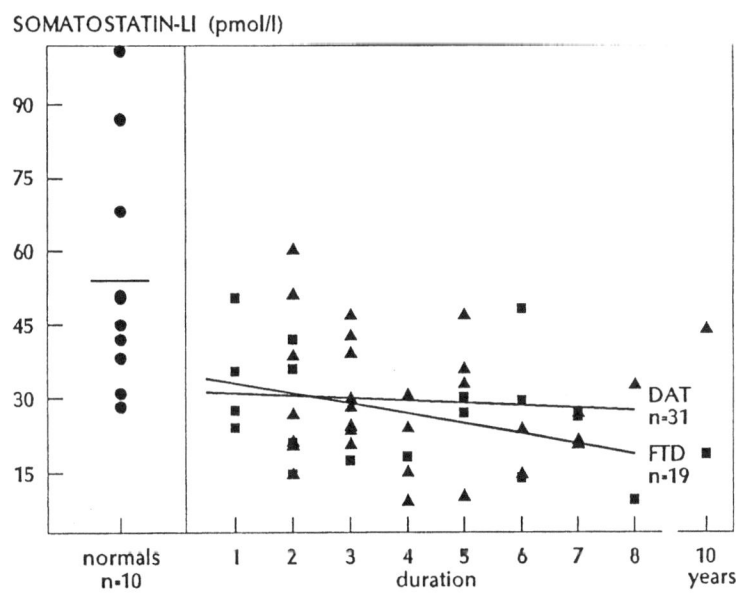

Fig. 4. The relationship between CSF somatostatin-LI and duration of dementia of Alzheimer type (DAT) and of frontotemporal degeneration (FTD)

for DSIP-LI level are shown in Table 3. DSIP has been identified as a sleep inducing hypnogenic factor, with other unknown effects in CNS. It is primarily associated with functions of the hypothalamic-pituitary-adrenal (HPA) axis and has been found to co-localize in ACTH/MSH cells in the hypophysis (Kafi et al., 1979; Monnier et al., 1977; Monnier et al., 1963; Schneider-Helmert et al., 1981; Marcus et al., 1986). It was noted that the level of DSIP-LI was reduced from 553 ± 133 pmol/l in controls to 375 ± 79 pmol/l in DAT ($p < 0.001$) (Fig. 5). In contrast, there was a tendency of an increase in DSIP-LI in FTD (773 ± 448), compared to healthy controls. The differences between DSIP-LI in DAT and FTD was significant ($p < 0.001$). DSIP-LI tended to increase with age in normals (Fig. 6) as well as in DAT and FTD. However, the relationship between DSIP-LI and duration of dementia seem to be different in DAT and FTD. In FTD there was a slight increase, while in DAT there was no such change (Fig. 7).

These observations in DAT and FTD are original ones since there is no comparable study reported in the literature. The reason for the observed difference in DSIP-LI of the two forms of dementia is unclear. Speculatively there might be lesions in systems controlling the release of central DSIP or, alternatively, another feedback system may be the primary lesion which influences the DSIP release. The neuro-endocrinological analysis in different types of dementia may explain some aspects of the clinical differences.

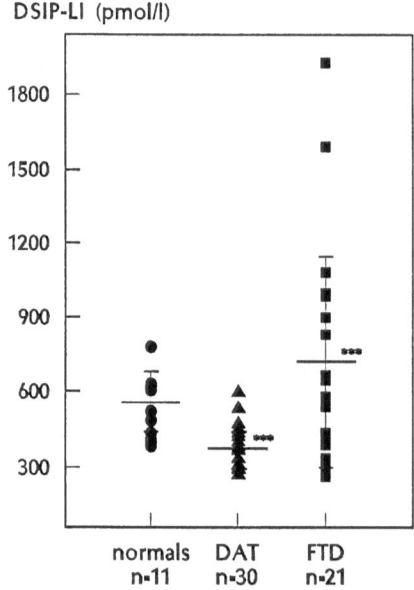

Fig. 5. Mean CSF levels of DSIP-LI in patients with dementia of Alzheimer type (DAT) and frontotemporal degeneration (FTD), and in normals. *** $p < 0.001$

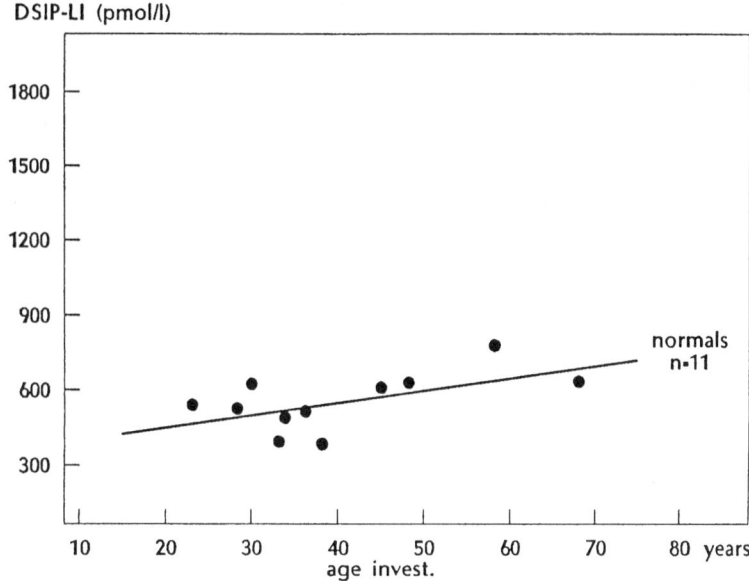

Fig. 6. The relationship between CSF DSIP-LI and age at investigation in normals

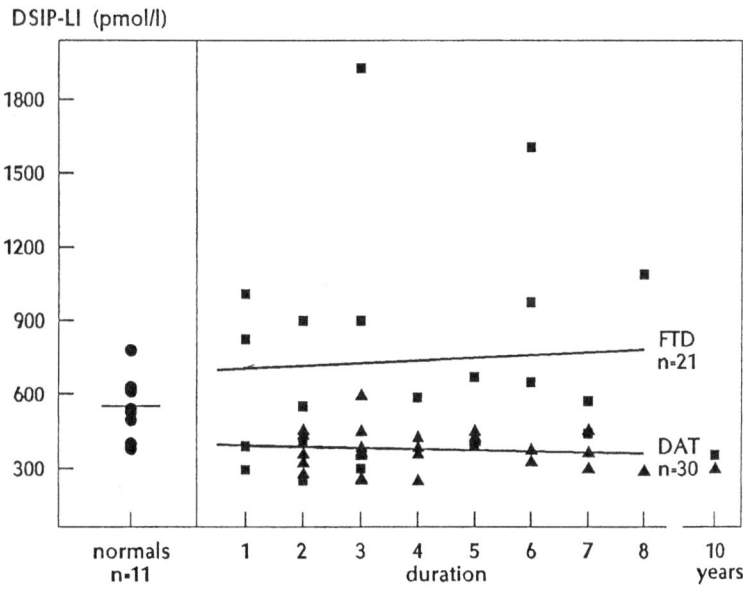

Fig. 7. The relationship between CSF DSIP-LI and duration of dementia of Alzheimer type (DAT), and of frontotemporal degeneration (FTD)

Conclusion

DAT and FTD show both similarities and differences in CSF levels of the three neuropeptides, S-LI, NPY-LI and DSIP-LI. This indicates the possibility of a neurochemical differentiation between two types of primary degenerative dementia. It may provide new tools for the recognition of dementia subgroups and of understanding the neurochemical disturbances behind primary degenerative disorders.

Acknowledgements

The authors wish to thank Siv Karlsson for producing the illustrations and Aniko Wolf for typing the manuscript. This study was supported by the Swedish Medical Research Council (projects 4969,3950) and the Greta and Johan Kock Foundation, Trelleborg.

References

Beal MF, Mazurek MF, McKee MA (1987) The regional distribution of somatostatin and neuropeptide Y in control and Alzheimer's disease striatum. Neurosci Lett 79:201–206

Brun A, Gustafson L (1976) Distribution of cerebral degeneration in Alzheimer's disease. A clinico-pathological study. Arch Psychiatr Nervenkr 223:15–33

Brun A, Englund E (1981) Regional pattern of degeneration in Alzheimer's disease: neuronal loss and histopathological grading. Histopathology 5:549–564

Brun A (1987) Frontal lobe degeneration of non-Alzheimer type. I. Neuropathology. Arch Gerontol Geriatr 6:193–208

Brun A (1987) Frontal lobe degeneration of non-Alzheimer type. I. Clinical picture and differential diagnosis. Arch Gerontol Geriatr 6:193–208

Brun A, Gustafson L (1988) Zerebrovaskuläre Erkrankungen. In: Kisker KP, et al (Hrsg) Psychiatrie der Gegenwart. Organische Psychosen. Springer, Berlin Heidelberg New York Tokyo, S 253–295

Davies P, Katzman R, Terry RD (1980) Reduced somatostatin-like immunoreactivity in cerebral cortex from cases of Alzheimer's disease and Alzheimer senile dementia. Nature 288:279–280

Ekman R, Larsson I, Malmquist M, Thorell J (1983) Radioimmunoassay of delta-sleep-inducing peptide using an iodinated p-hydroxyphenylpropionic acid derivative as tracer. Regul Pept 6:371–378

Ekman R, Wahlestedt C, Böttcher G, Sundler F, Hakanson R, Panula P (1986) Peptide YY-like immunoreactivity in the central nervous system of the rat. Regul Pept 16:157–168

Gustafson L, Risberg J, Johanson M, et al (1984) Evaluation of organic dementia by regional cerebral blood flow measurements and clinical and psychometric methods. In: Fieschi C, et al (eds) Effects of aging on regulation of cerebral blood flow and metabolism. Monogr Neural Sci 11:111–117

Gustafson L (1987) Frontal lobe degeneration of non-Alzheimer type. I. Clinical picture and differential diagnosis. Arch Gerontol Geriatr 6:209–223

Gustafson L, Brun A (1989) The Scandinavian view. In: Hovaguimian T, et al (eds) Classification and diagnosis of Alzheimer disease: an international perspective. Hogrefe and Huber, Toronto Lewiston NY Bern Stuttgart Göttingen, pp 17–23

Gustafson L, Nilsson L (1982) Differential diagnosis in presenile dementia on clinical grounds. Acta Psychiatr Scand 65:194–209

Hachinski VC, Iliff LD, Zilka E, et al (1975) Cerebral blood flow in dementia. Arch Neurol 32:632–637

Kafi S, Monnier M, Gaillard J-M (1979) The delta sleep inducing peptide (DSIP) increase duration of sleep in rats. Neurosci Lett 13:169–172

Marcus V, Graf A, Kastin A (1986) Delta-sleep-inducing peptide (DSIP): an update. Peptides 7:1165–1187

Monnier M, Koller T, Graber S (1963) Humoral influence of induced sleep and arousal upon electrical brain activity of animals with cross circulation. Exp Neurol 8:264–277

Monnier M, Dudler L, Gächter R, Schoenenberger G (1977) Delta sleep inducing peptide (DSIP). EEG and motor activity in rabbits following intravenous administration. Neurosci Lett 5:9–13

Nakamura S, Vincent SR (1986) Somatostatin- and neuropeptide Y-immunoreactive neurons in the neocortex in senile dementia of Alzheimer's type. Brain Res 370:11–20

Risberg J (1980) Regional cerebral blood flow measurements by 133 Xe inhalation: methodology and applications in neuropsychiatry and psychiatry. Brain Lang 9:9–34

Risberg J, Gustafson L (1982) 133 Xe cerebral blood blow in dementia and in neuropsychiatry research. In: Magistretti PL (ed) Functional radionuclide imaging of the brain. Raven Press, New York, pp 151–158

Risberg J (1987) Frontal lobe degeneration of non-Alzheimer type. III. Regional cerebral blood flow. Arch Gerontol Geriatr 6:225–233

Rossor MN, Ellison PC, Mountjoy CQ, et al (1980) Reduced amounts of immunoreactive somatostatin in the temporal cortex in senile dementia of Alzheimer type. Neurosci Lett 20:373–377

Schneider-Helmert D, Gnirss F, Monnier M, Schenker J, Schoenenberger G (1981) Acute and delayed effect of DSIP (delta sleep inducing peptide) on human sleep behaviour. Int J Clin Pharmacol Ther Toxicol 19:341–345

Wallengren J, Ekman R, Sundler F (1987) Occurrence and distribution of neuropeptides in the human skin. Acta Derm Venereol 67:551–558

Widerlöv E, Lindström H, Wahlestedt C, Ekman R (1988) Neuropeptide Y and peptide YY as possible cerebrospinal fluid markers for major depression and schizophrenia, respectively. J Psychiatr Res 22(1):69–79

Authors' address: L. Minthon, M. D., Department of Psychogeriatrics, University of Lund, P.O. Box 638, S-22006 Lund, Sweden.

J Neural Transm (1990) [Suppl] 30: 69–78

Changes in signal transduction in Alzheimer's disease

**S. Shimohama[1], H. Ninomiya[1], T. Saitoh[2], R. D. Terry[2], R. Fukunaga[3],
T. Taniguchi[3], M. Fujiwara[4], J. Kimura[1], and M. Kameyama[5]**

[1] Department of Neurology, Faculty of Medicine, Kyoto University, Japan
[2] Department of Neurosciences, School of Medicine, University of California,
San Diego, USA
[3] Department of Neurobiology, Kyoto Pharmaceutical University,
[4] Department of Pharmacology, Faculty of Medicine, Kyoto University, and
[5] Department of Neurology, Sumitomo Hospital, Japan

Summary. We studied the signal transduction system including the receptor and protein kinase C (PKC) in Alzheimer's disease (AD) brains. We used ^{3}H-TCP as a ligand for the NMDA receptor-ion channel complex. The total concentrations of ^{3}H-TCP binding sites were significantly reduced in AD frontal cortex. ^{3}H-TCP binding sites spared in AD brains retained the affinity for the ligand and the reactivity to NMDA, L-glutamate, and glycine. We utilized antibodies to assess the degree of involvement of different PKC isoforms in AD. The concentration of PKC (βII) was lower in AD particulate fractions and higher in AD cytosol fractions. Immuno-cytochemical studies revealed reduced numbers of anti-PKC (βII)-immunopositive neurons. Anti-PKC (α) faintly stained entire plaques and surrounding glial cells. Anti-PKC (βI) stained dystrophic plaque neurites. Anti-PKC (βII) stained the amyloid-containing portions of plaques. These results suggest an involvement of second messenger cascades in the pathogenesis of AD in addition to neurotransmitters and their receptors.

Introduction

Alzheimer's disease (AD) is a neurodegenerative disease characterized by progressive dementia. The etiology of AD is not known. In the 1970's, it was clearly established that AD causes a serious decrease in the neurotransmitter acetylcholine (ACh), one of several chemical messengers in the brain, and one that is closely tied to memory and other mental functions. This finding stimulated considerable scientific activity aimed at identifying the exact nature of the defect and finding a way to treat the disease with drugs, in much the same way as Parkinson's disease treated. However, AD now appears to be too complex a problem, involving many different elements of brain chemistry and anatomy, for this approach.

One of the characteristics of AD is the selective loss of large neurons in the cerebral cortex. Several neurotransmitters, their marker enzymes, and certain neurotransmitter receptors are reduced in brain tissue from Alzheimer patients. These include ACh, samotostatin, norepinephrine, serotonin, substance P, glutamate, corticotropin releasing factor, and neuropeptide Y. Some of these neurotransmitters have been co-localized with neuritic plaques and neurofibrillary tangle-bearing neurons, well documented morphological abnormalities found in Alzheimer brains. It may be, therefore, that the large neurons containing these neurotransmitters and receptors are the ones dying in the Alzheimer cortex. Selective deficits in neurotransmitter metabolism, however, are not likely to cause cell death. No biochemical deficits have yet been reported to explain this neuronal cell death.

The NMDA receptor, well characterized as a subtype of excitatory amino acid receptor in the mammalian CNS, is a focus of current interest for its possible role in learning and memory (Collingridge and Bliss, 1987), and in neuronal cell death under various pathological conditions (Rothman and Olney, 1987). Autoradiographic studies using AD postmortem brains have shown a reduction in ^{3}H-TCP binding or in N-methyl-D-aspartate (NMDA)-sensitive ^{3}H-glutamate binding in hippocampal sections (Geddes et al., 1986; Greenamyre et al., 1987; Monaghan et al., 1987; Maragos et al., 1987). N-(1-(2-thienyl)cyclohexyl)3,4-piperidine (TCP) is a derivative of an anesthetic, phencyclidine. Recent biochemical studies have shown that at least part of the recognition sites are located within ion channels coupled to NMDA receptors (Kemp et al., 1987). More recently, Simpson et al. (1988) reported reduced ^{3}H-TCP binding in frontal cortex from AD brains, although the possible changes in kinetic parameters were not examined in that study. These findings suggested a loss of NMDA receptor-ion channel complex in AD brains.

Receptor-mediated polyphosphoinositide (PPI) metabolism is now generally accepted as ubiquitous transmembrane signalling mechanism by which many Ca^{2+}-mobilizing agonists, including hormones, neurotransmitters and growth factors, activate their target cells. The main reaction in receptor-stimulated PPI turnover is a GTP-binding protein-mediated hydrolysis of phosphatidylinositol-(4,5)-biphosphate (PIP_2) by phospholipase C, resulting in at least two second messengers, inositol-(1,4,5)-triphosphate (IP_3) and 1,2-diacylglycerol (DG). While DG activates the Ca^{2+}- and phospholipid-dependent protein kinase C (PKC), IP_3 stimulates the release of intracellular Ca^{2+} from a non-mitochondrial store, presumably endoplasmic reticulum.

PKC is a family of closely related enzymes, which is involved in many cellular functions including adhesion, secretion, growth, and differentiation. In the CNS, PKC appears to be involved in several important specialized functions. First, the efficiency of neurotransmission seems to be controlled by the degree of phosphorylation of proteins, which is, in turn, controlled by PKC. In fact, PKC activation is a critical part of the process of long term potentiation. Second, PKC is involved in the survival of neurons. Many neuronal trophic factors exert their function of supporting neuronal survival through the activation of PKC (Nishizuka, 1988, 1989).

Based on these informations, we set out to know whether the changes of the signal transduction system contribute to the pathogenesis of AD.

First, we examined the changes of ^{3}H-TCP binding in AD brains. Extensively washed membrane preparations were used to minimize the possible effects of endogenous substances (Foster and Wong, 1987; Ransom and Stec, 1988). We examined the effects of NMDA receptor agonists on ^{3}H-TCP binding and the correlation between ^{3}H-TCP and NMDA-sensitive ^{3}H-glutamate binding sites in the same membrane preparations.

Second, we examined the changes of PKC in AD brains. Brain contains several types of PKC isozymes. It is of interest to analyze these isozymes of PKC in AD brains, because some of them are relatively restricted to brain tissue and may be involved in specialized functions. The region-specific and selective subcellular distribution of the different isozymes suggests selective functions. In this study, we employed antibodies to four isozymes of PKC to study the PKC abnormality in AD brain. Antisera were raised in rabbits against synthetic peptides predicted from the human cDNA sequence corresponding C-terminal portion of four PKC isozymes. For quantification of the Mr 80 000 PKC band, we used the Western blot analysis. To know if altered PKC biochemistry is restricted to certain cell types or is a general phenomenom, we used immunocytochemical method.

These studies will clearly verify our present hypothesis of the possible deficiency in receptors and intracellular signal transduction system in AD. At the same time, they will provide us a better understanding of the biochemical basis of the pathophysiology of the disease.

Materials and methods

I. ^{3}H-TCP binding

^{3}H-TCP (60 Ci/mmol) and L-(3,4-^{3}H) glutamate (69.7 Ci/mmol) were purchased from New England Nuclear (Boston, MA, USA).

Brain tissues were obtained at autopsy from five patients (80 ± 3 yr) diagnosed clinically and histopathologically as having AD, and from five subjects (79 ± 2 yr) with no clinical or morphological evidence of brain pathology. The groups were matched for age and time to autopsy. Immediately after autopsy, brains were halved sagittally and one-half was examined histopathologically. AD was diagnosed histopathologically from the widespread presence of neuronal cell loss, senile plaques and neurofibrillary tangles in the neocortex and hippocampus. We examined the frontal cortex.

Membranes were prepared according to the method of Ransom and Stec (1988) with a slight modification. ^{3}H-TCP binding assays were performed according to the method of Ransom and Stec (1988) with a slight modification. ^{3}H-Glutamate binding assays were performed using Triton X-100-treated membrane preparations according to the method modified from Ogita and Yoneda (1988).

Results are given as means $\pm$ SEM values from n experiments. The total concentration (B_{max}) of the binding sites and the apparent dissociation constant (K_d) were determined from computer assisted linear regression analysis of Scatchard plots. Statistical analysis of binding data was done by ANOVA.

II. PKC isozymes

A total of 12 brains were studied. There were six cases of AD (75 ± 5 yr) and six age-matched normal controls (70 ± 10 yr). All the AD patients were clinically demented, whereas those in the control group were not. Following removal, the brains were divided sagittally and the left hemibrain was fixed in 10% formalin while the right hemibrain was frozen at $-70\,°C$.

Dissected cortex was placed in 10 volumes of the homogenization buffer (0.32 M sucrose; 5 mM HEPES, pH 8.0; 5 mM benzamidine; 2 mM mercaptoethanol; 3 mM EGTA; 0.5 mM $MgSO_4$; 10 µM sodium meta-vanadate; 0.1 mM PMSF; 10 µg/ml leupeptin; 5 µg/ml pepstatin A and 10 µg/ml aprotinin) and homogenized by two 5 sec strokes of a Polytron homogenizer.

This homogenate was centrifuged 1 hr at $100,000 \times g$ at $2\,°C$ to separate the cytosol from particulate fraction. Both fractions were adjusted to 1 mg protein/ml by adding the homogenization buffer, divided into aliquots, frozen in an ethanol/dry-ice bath, and stored at $-70\,°C$ until used for Western blot analysis.

After 10 to 14 days of formalin fixation, left hemibrains were examined grossly, dissected, and blocks were taken from several regions. Hematoxylin and eosin (H & E) and Nissl preparations were examined for overall neuropathological evaluation. Thioflavin S-stained sections were viewed with ultraviolet illumination and fluorescein filters for the identification of neocortical neurofibrillary tangles and neuritic plaques. All of AD brains had large numbers of neuritic plaques and neurofibrillary tangles in the neocortex.

Antisera were raised in rabbits against synthetic peptides predicted from the human cDNA sequence corresponding C-terminal portion of four PKC isozymes, α, βI, βII and γ. The sequences used were PQFVHPILQSAV (PKC (α), 661–672), SYTNPEFVINV (PKC (βI), 661–671), NSEFLKPEVKS (PKC (βII), 663–673) and SPISPTVPVM (PKC (γ), 687–697). The C-terminal cysteine was added to the peptides so that they can be conjugated to proteins by the method of Green et al. (1982). All the antisera gave a titer between 1/2000 and 1/5000 on dot blots with the individual peptides. Antisera were further purified by affinity chromatography on immobilized antigen columns prepared by coupling free peptide antigens to activated aldehyde-agarose (Actigel A, Sterogene, Arcadia, Calif.). The staining on blots was abolished by the inclusion of 20 µg/ml of the free corresponding peptide demonstrating the specificity of our antisera.

Protein blot and immunological detection of PKC were performed as follows. Protein (50 µg) in Laemmli sample buffer (Laemmli, 1970) was electrophoresed on

6.5%–16% SDS polyacrylamide gels, blotted to nitrocellulose, blocked with PBS containing 0.1% Tween 20 (TPBS) for 12 h, and incubated with primary antibody (1/2000 dilution in PBS containing 3% BSA) 1 h at 4 °C. Blots were then washed with TPBS, incubated with 0.5 μCi/ml iodinated protein A in PBS containing 3% BSA, and then washed and autoradiographed on Kodak X-omat RP film at −70 °C. Films were then developed with Konica film developer and scanned with an LKB densitometer for quantification of the Mr 80000 PKC band.

The immunocytochemical methods were essentially as described by Cole et al. (1989). To show the specificity of the staining, control and AD sections were incubated overnight with antiserum absorbed with 4 mg/ml of the free peptide and without primary antiserum.

Results

I. ^{3}H-TCP binding

Schachard plots of specific ^{3}H-TCP binding to either extensively washed, or unwashed crude cortical membrane preparations were linear, suggesting the presence of a homogenous binding site. Membrane washing resulted in an apparent decrease in the affinity of ^{3}H-TCP for its binding site, with no change in the maximal binding capacity for washed and unwashed membranes, respectively. In the thoroughly washed membrane preparations, either L-glutamate or NMDA markedly enhanced ^{3}H-TCP binding. Addition of glycine, which in itself has little effect on ^{3}H-TCP binding, produced a significant increase in L-glutamate or NMDA-induced enhancement of the binding. The presence of either substance caused no change in B_{max} values and the enhancement of the binding was resulted solely from an increase in the affinity of the ligand.

In AD membrane preparations, there was a significant reduction in B_{max} values without any change in K_d values (Table 1). Correlation between ^{3}H-TCP and NMDA-sensitive ^{3}H-glutamate binding sites was determined using the same membrane preparations preincubated with 0.08% Triton X-100. Triton X-100 treatment caused no significant change in either B_{max} or K_d values of ^{3}H-TCP binding. The reduction in ^{3}H-TCP binding capacity was also apparent in Triton X-100-treated membrane

Table 1. Decreased [^{3}H]TCP binding in AD frontal cortex

	K_d (nM)	B_{max} (fmol/mg protein)
Control (n = 5)	44 ± 8	456 ± 25
AD (n = 5)	38 ± 4	214 ± 14*

* p < 0.01

preparations, with a linear correlation between the concentrations of ^{3}H-TCP binding sites and NMDA-sensitive ^{3}H-glutamate binding capacity. Actual NMDA-sensitive ^{3}H-glutamate binding capacity (determined at 20 nM ^{3}H-glutamate) was significantly decreased in AD group (Table 2). We examined the enhancement of ^{3}H-TCP binding by L-glutamate, NMDA, and glycine in control and AD membrane preparations. The effects of these substances showed no significant change between control and AD membrane preparations.

Statistical analysis by ANOVA revealed no significant correlations between age, sex or postmortem delay and binding data in the control group, or between binding data and duration of illness in the AD group.

II. PKC isozymes

The specificity of the antisera used for this study was tested on Western blots where total particulate proteins or total soluble proteins were electro-transferred onto the nitrocellulose paper. Rabbit antibodies against C-terminal synthetic peptide of four PKC isozymes α, βI, βII, and γ all recognized the PKC as a Mr 80000 protein in both particulate and soluble fractions. The staining of this band was eliminated by incubation of each antibody with excess peptides showing the specificity.

Anti-PKC (βII) showed a reduced staining on Western blots in the particulate fraction from frontal cortex with a concomitant increase in the soluble fraction. Anti-PKC (βI) showed an elevated staining in the soluble fraction. Anti-PKC (α) and -(γ) staining were not altered in AD cortex (Table 3).

In control neocortical sections, anti-PKC (α) and -(βI) moderately stained the cytoplasm of the pyramidal cells. Some immunoreactivity was noted in small neurons, glial cells, and small vessel walls. In AD neocortical sections no significant changes in the patterns of pyramidal neuron staining were observed with anti-PKC (α) and -(βI). Some immunoreactivity was also noted in nuclei, neurites, small neurons, and walls of small vessels. While total large neuron populations in layers 3 and 5 were reduced by half in AD, anti-PKC (βII) immunoreactive large neurons were reduced by

Table 2. Decreased NMDA-sensitive [^{3}H]glutamate binding in AD frontal cortex

	[^{3}H]glutamate bound (fmol/mg protein)
Control (n = 5)	132 ± 16
AD (n = 5)	53 ± 8*

* p < 0.01

Table 3. Immunoquantification of PKC isozymes in AD frontal cortex

	Particulate fraction	Cytosol fraction
PKC-α	→	→
-βI	→	↑
-βII	↓	↑
-γ	→	→

The Mr 80000 PKC band on autoradiographs was scanned with an LKB ultrascan densitometer and the relative densities from AD and control samples were compared by Student's t-test analysis. ↓: Significantly decreased $p < 0.05$; ↑: Significantly increased $p < 0.05$

80%, and those remaining stained less intensely than the controls. Anti-PKC (α) reacted diffusely with entire plaques and with glial cells around them. With anti-PKC (βII) an intense fibrillar immunoreactivity was found in the amyloid-containing zones of plaques. About 80% of the thioflavin-S-positive classical plaques were moderately immunoreactive to anti-PKC (βII) in the cores and fibrillar zones. Anti-PKC (βI) stained dystrophic plaque neurites. Anti-PKC (γ) did not stain plaques. In all cases, staining could be absorbed out with an excess of competing free peptide immunogens.

Discussion

The present study revealed that the reduction in ³H-TCP binding is due to a decrease in the total number of the binding sites. There was a linear correlation between the number of ³H-TCP binding sites and that of NMDA-sensitive ³H-glutamate binding sites. These results indicate that the primary change in NMDA receptor-ion channel complex in AD brains is the reduction of its number, possibly secondary to loss of neurons bearing these receptor-ion channel complexes.

We have previously reported regional changes in nicotinic and muscarinic cholinergic, α- and β-adrenergic, and benzodiazepine receptors in AD brains (Shimohama et al., 1986 a, b, 1987, 1988), and the reduction in number in frontal cortex was recorded only for benzodiazepine receptors. These results suggest the specificity of the reduction in NMDA receptor-ion channel complexes in AD brains, providing a supportive evidence to the hypothesis that excitatory amino acid receptors may play a role in the pathogenesis of AD (De Boni and MacLachlan, 1985; Choi, 1988).

There was no significant difference in K_d values of ³H-TCP binding between control and AD membrane preparations and they also retained the reactivity to NMDA agonists and glycine. These results suggest that the functional linkage within the receptor complex spared in AD brains remains normal. If we assume that the neuronal loss in AD mainly results from the excitotoxity caused by the activation of NMDA receptor-ion channel complex, one possible explanation for these results is that the remaining receptor complex in AD brains is located on surviving neurons which are not involved in the pathogenesis of AD.

Reduced PKC levels determined by radioactive phorbol ester binding in particulate fractions from AD cortex and a trend toward elevated PKC levels in AD soluble fractions are previously reported in AD brains (Cole et al., 1988). Although PKC is a family of closely related kinases, the individual isozymes may have distinct distributions and functions (Nishizuka, 1988, 1989). It is also not known if altered PKC biochemistry is restricted to certain cell types or is a general phenomenon. The present study answers both of these questions.

The reduction in PKC(βII) was detected by both biochemical and immunocytochemical methods in the cortex. There was a general trend toward decreased particulate and increased supernatant or cytosolic immunostaining in AD on Western blots. The most significant changes in the cortex were the reduction in anti-PKC (βII) immunoreactivity in the pellet and the concomitant increase in the supernatant observed in Western blots. Consistent with these biochemical data, a significant decrease in anti-PKC (βII) immunoreactivity was observed in AD cortical sections. This is attributable to both preferential loss of PKC (βII) immunoreactive subpopulations of large neurons, and to reduced PKC (βII) content of those remaining. The cerebral cortex has very high PKC (βII) levels which account for the majority of PKC (Ase et al., 1988), suggesting that changes in PKC (βII) largely account for the reduced PKC activity and amount in the pellet fractions which were reported previously (Cole et al., 1988).

In AD brains, there are two prominent pathological changes, neuritic plaques and neurofibrillary tangles. Anti-PKC (βII) antibody immunolabeled material colocalized with amyloid in the cores of plaques. This labeling was absorbed out by peptide. Anti-PKC (βII) staining of plaques is similar to anti-amyloid-protein staining. The material stained with anti-PKC (βII) is not stained with anti-PKC (α) or -(βI). It is therefore likely that amyloid staining of neuritic plaques with anti-PKC (βII) antibody is specific. PKC has been shown to phosphorylate amyloid precursor protein in vitro (Gandy et al., 1988) indicating one hypothetical connection between PKC and amyloid precursor processing. The neuritic component of plaques was stained with anti-PKC (βI) antiserum, but not with antisera against PKC (α), -(βII), or -(γ). We have observed anti-PKC (α) staining of reactive glia in the rat hippocampus with experimental lesions (Shomohama et al., 1988). The antiserum against the PKC (γ) isoform was negative with glia.

These results present evidence for alterations of PKC isozymes in AD. While the cause of these alterations remains unknown, they may not simply reflect the neuropathology of AD since we have found similar PKC deficits in AD fibroblasts (Huynh et al., 1989). The changes in immunocytochemical staining observed in AD tissue sections and with homogenates on blots may reflect, in part, reduced activation and diminished translocation of PKC in AD, which could be directly related to neuronal dysfunction. The aberrant PKC behavior reported here should be explored further in hopes of identifying an underlying cause for neuronal dysfunction and death in AD.

Moreover, both NMDA receptor-ion channel couple and PKC are closely related to intracellular Ca^{2+} homeostasis. Altered Ca^{2+} homeostasis may be pathophysiologically important in AD. Deficits in Ca^{2+} movements across membranes may lead to the formation and phosphorylation of abnormal proteins during AD. Proteolysis and axonal transport are depressed by decreased Ca^{2+}, thus, the age-related reductions in Ca^{2+} uptake may lead to the accumulation of abnormal proteins in neuronal cell bodies.

In conclusion, the present study suggests an involvement of second messenger cascades in the pathogenesis of AD, in addition to neurotransmitters and their receptors.

References

Ase K, Saitoh N, Shearman MS, Kikkawa U, Ono Y, Igarashi K, Tanaka C, Nishizuka Y (1988) Distinct cellular expression of βI- and βII-subspecies of protein kinase C in rat cerebellum. J Neurosci 8:3850–3856

Choi DW (1988) Glutamate neurotoxicity and diseases of the nervous system. Neuron 1:623–634

Cole G, Dobkins LA, Hansen LA, Terry RD, Saitoh T (1988) Decreased levels of protein kinase C in Alzheimer brain. Brain Res 452:165–170

Cole G, Masliah E, Huynh TV, DeTeresa R, Terry RD, Okuda C, Saitoh T (1989) An antiserum against amyloid β-protein precursor detects a unique peptide in Alzheimer brain. Neurosci Lett 100:340–346

Collingridge GL, Bliss TVP (1987) NMDA receptor – their role in long-term potentiation. TINS 10:288–293

De Boni U, McLachlan DR (1985) Controlled induction of paired helical filaments of the Alzheimer type in cultured human neurons, by glutamate and aspartate. J Neurol Sci 68:105–118

Foster AC, Wong EHF (1987) The novel anticonvulsant MK-801 binds to the activated state of the N-methyl-D-aspartate receptor in rat brain. Br J Pharmacol 91:403–409

Gandy S, Czernik AN, Greengard P (1988) Phosphorylation of Alzheimer disease amyloid precursor peptide by protein kinase C and Ca^{2+}/calmodulin-dependent protein kinase II. Proc Natl Acad Sci USA 85:6218–6221

Geddes JW, Chang-Chui H, Cooper SM, Lott IT, Cotman CW (1986) Density and distribution of NMDA receptors in the human hippocampus in Alzheimer's disease. Brain Res 399:156–161

Green N, Alexander H, Olson A, Alexander S, Shinnick TM, Sutcliffe JG, Lerner RA (1982) Immunogenic structure of the influenza virus hemagglutinin. Cell 28:477–487

Greenamyre JT, Penney JB, D'Amato CJ, Young AB (1987) Dementia of the Alzheimer's type: changes in hippocampal L-^{3}H-glutamate binding. J Neurochem 48:543–551

Huynh TV, Cole G, Katzman R, Huang K-P, Saitoh T (1989) Reduced PKC immunoreactivity and altered protein phosphorylation in Alzheimer's disease fibroblasts. Arch Neurol 46:1195–1199

Kemp JA, Foster AC, Wong HF (1987) Non-competitive antagonists of excitatory amino acid receptors. TINS 10:294–298

Laemmli UK (1970) Cleavage of structural proteins during the assembly of the head of bacteriophage T4. Nature 227:680–685

Maragos WF, Chu DCM, Young AB, D'Amato CJ, Penney Jr B (1987) Loss of hippocampal ^{3}H-TCP binding in Alzheimer's disease. Neurosci Lett 74:371–376

Monaghan DT, Geddes JW, Yao D, Chung C, Cotman CW (1987) ^{3}H-TCP binding sites in Alzheimer's disease. Neurosci Lett 73:197–200

Nishizuka Y (1988) The molecular heterogeneity of protein kinase C and its implications for cellular regulation. Nature 334:661–665

Nishizuka Y (1989) Studies and prospectives of the protein kinase C family for cellular regulation. Cancer 63:1892–1903

Ogita K, Yoneda Y (1988) Disclosure by Triton X-100 of NMDA-sensitive ^{3}H-glutamate binding sites in brain synaptic membranes. Biochem Biophys Res Commun 153:510–517

Ransom RW, Stec NL (1988) Cooperative modulation of ^{3}H-MK-801 binding to the N-methyl-D-aspartate receptor-ion channel complex by L-glutamate, glycine, and polyamines. J Neurochem 51:830–836

Rothman SM, Olney JW (1987) Excitotoxicity and the NMDA receptor. TINS 10:299–302

Shimohama S, Taniguchi T, Fujiwara M, Kameyama M (1986a) Changes in nicotinic and muscarinic cholinergic receptors in Alzheimer-type dementia. J Neurochem 46:288–293

Shimohama S, Taniguchi T, Fujiwara M, Kameyama M (1986b) Biochemical characterization of α-adrenergic receptors in human brain and changes in Alzheimer-type dementia. J Neurochem 47:1294–1301

Shimohama S, Taniguchi T, Fujiwara M, Kameyama M (1987) Changes in β-adrenergic receptor subtypes in Alzheimer-type dementia. J Neurochem 48:1215–1221

Shimohama S, Taniguchi T, Fujiwara M, Kameyama M (1988) Changes in benzodiazepine receptors in Alzheimer-type dementia. Ann Neurol 23:404–406

Shimohama S, Saitoh T, Gage FH (1988) Protein kinase C in hippocampus and septum following fimbria-fornix transection. Soc Neurosci Abstr 14:19

Simpson MDC, Royston MC, Deakin JFW, Cross AJ, Mann DMA, Slater P (1988) Regional changes in ^{3}H-D-aspartate and ^{3}H-TCP binding sites in Alzheimer's disease brains. Brain Res 462:76–82

Authors' address: Dr. S. Shimohama, Department of Neurology, Faculty of Medicine, Kyoto University, 54 Shogoinkawaharacho, Sakyo-ku, Kyoto 606, Japan.

Subject Index

acetylcholine (ACh) 1, 2, 4–10, 25–31, 69, 70
acetylcholinesterase (AChE) 3, 13, 15, 17–20, 25, 36
– inhibitor 14
acetyl CoA 31
acid phosphatase 17, 20
ACTH/MSH 64
age 28, 29, 35, 37, 40, 52, 60, 62–65
alkaline protease 15
Alzheimer 16
– dementia 33
– 's disease (AD) 1, 13, 14, 18, 38, 40, 41, 53, 69–77
– – /SDAT 37
– type dementia (DAT) 25, 36, 39
amphetamine 9
amyloid 69, 75, 76
– core 21
α and β-adrenergic 75
α_1-antichymotrypsin 21
antimucarinic 75
anti-PKC(α) 76
– – (β_1) 76
arginine vasopressin (AVP) 33, 36, 40, 41
arousal 34
aspartate 36

basal forebrain 8
behavioral testing 2
benzodiazepine receptor 75
Binswanger's disease (BD) 45–53
blood-brain barrier 9, 10
B_{max} 73
brain dialysis 1, 3
– stem 35
butyrylcholinesterase 13, 15, 18–20
BW284c51 15

cerebral hemisphere 53
cerebrospinal fluid (CSF) 25–31, 35–40, 45–53, 57–66

choline 4, 5, 25–31
– acetyl transferase (CAT, ChAT) 1, 13, 25, 31, 36
– esterase 31
– oxidase 3
cingulate cortex 37
citalopram 41
collagen 13, 21
collagenase 13, 15, 18, 20, 21
congo red 15, 16, 18
cortex 36
corticotropin-releasing factor (CRF) 36, 40, 70
cytosol (cytoplasm) 74–76

delta sleep inducing peptide (DSIP) 57–60, 62, 64–66
dementia of the Alzheimer type (DAT) 26, 30, 31, 34, 45–52, 57, 58, 60–66
– with frontotemporal degeneration of non-Alzheimer type (FTD) 57–66
demyelination 52
dexamethasone suppression test (DST) 40, 41
1,2-diacylglycerol (DG) 70
dopamine (DA) 36, 53
– beta-hydroxylase 36
duration 60, 62–65

eating disorder 40
electrochemical detector 3, 25
β-endorphin (β-Ep) 45, 46, 48, 50–53
ethylhomocholine (EHC) 4, 5, 26, 27
excitotoxity 75

frontal cortex 69, 72, 75
FTD 57–66

GABA 36
– receptor 36
galanin 40, 41
glia 76

glial cell 69, 74
gliosis 38
globus pallidus 37
glutamate 69, 73, 74
gyrus cinguli 39

high-performance liquid chromatography (HPLC) 3, 9, 25, 26, 48, 59
hippocampus 8, 36, 41, 72, 76
homovanillic acid (HVA) 36, 45–49, 53
5-hydroxyindoleacetic acid 33–36, 38, 40, 41, 45–48, 50, 53
hydroxymethylphenylglycerol (HMPG) 36
5-hydroxytryptamine 33–36, 38, 40, 41
5-hydroxytryptophan 34, 35
hyperactivity 9
hypothalamus 33, 36, 40, 41, 52, 53
– pituitary-adrenal (HPA) axis 40, 64
hypoxia 52, 53

imipramine 33, 37
– binding 36
inositol-(1,4,5)-triphosphate (IP$_3$) 70
ischemic score 51, 52, 58

Kd 73, 75

liquid cation-exchange 26
locus caeruleus 37
lumbar puncture 26

memory 31
methylscopolamine 1, 2, 4, 7–10
microsomes 14
mini-mental state (MMS) score 46, 48, 49, 51–53
mitochondria 14, 20
monoamine oxidase (MAO) 37
– – B 36–39
MRI 45, 52
multi-infarct dementia (MID) 47
muscarinic cholinergic 75
– receptor 36
myelin fragment 14

neocortex 72
nerve ending 20
– – fraction 14
– – particle 16

neuritic plaque 70, 76
neurofibrillary tangle 19, 25, 70, 72, 76
neuropeptide 41, 59
– Y (NPY) 57–59, 60–62, 66, 70
neurotransmitter 70
nicotinic 75
– receptor 36
N-methyl-D-aspartate (NMDA) 69–71, 73–76
noradrenaline (NA) 36
norepinephrine 70
nuclei 14
2'-nucleotidase 17, 20
5'-nucleotidase 17, 20

opioid receptor 53

Parkinson's disease 31, 70
particulate 75, 76
passive avoidance task 1, 2, 4, 9
peptide YY 59
periventricular hyperintensity (PVH) 46, 48, 52
pernicious anaemia 38
phosphatidylinositol-(4,5)-biphosphate (PIP$_2$) 70
phospholipase C 70
phosphorylation 77
physostigmine 3, 9, 26
Pick's disease 57
plaque 69, 72, 75
platelet aggregation 38
polyphosphoinositide (PPI) 70
postsynaptic receptor 8
presynaptic autoreceptor 8
probenecid 33, 38
protease 13, 18
protein kinase C (PKC) 69–71, 75, 76
– – isozyme 71, 72, 74, 75, 77
– – (α) 69, 71, 74, 75
– – (β_I) 72, 74, 75
– – (β_{II}) 69, 72, 74–76
– – (γ) 76
putamen 37
pyramidal cell 74

radioimmunoassay 48, 59, 60
raphe nucleus 33, 37
receptor 69, 75
regional cerebral blood flow (rCBF) 58
retention trial 5

scopolamine 1, 2, 4, 5, 7–10
score 60, 62, 63
sedimentation coefficient 19
senile dementia of Alzheimer type
   (SDAT) 9
– plaque 13–16, 18–21, 25
serotonin 70
signal transduction 69, 71
sleep 34
– disorder 40
small vessel wall 74
sodium laurylsulfate 15
– 1-octanesulfonic acid 4
soluble fraction 14
somatostatin 36, 40, 41, 57–59, 62,
   63, 66, 70
spinal cord 53
striatum 3, 5, 8
subcellular distribution 13, 14, 16, 17
– – of AChE 16

substance P 70
substantia nigra 37
succinic dehydrogenase 17, 20
sucrose density gradient 15, 19
sympathetic nervous activity 34

tetramethylammonium 4
N-(1-(2-thienyl)cyclohexyl)3,4-piperi-
   dine (TCP) 69–75
trypsin 15
tryptophan 34, 35, 40
– hydroxylase 33, 37
tyrosine hydroxylase (TH) 36

vascular dementia 34, 58
vitamin B_{12} 38

Western blot analysis 71
white matter 39

Springer-Verlag Wien New York

Walther Birkmayer, Peter Riederer

Understanding the Neurotransmitters: Key to the Workings of the Brain

Translated from the German by Karl Blau

1989. 14 figures. XIX, 137 pages.
Soft cover DM 38,–, öS 266,–
ISBN 3-211-82100-7

This book demonstrates for the first time the connection between age, disease in old age, psychiatric disorders, pain, psychosomatic phenomena on the one hand and the function of neurotransmitters on the other and attempts to explain the significance of these substances for our behaviour. The authors therefore offer a biological approach to psychotherapy, drug dependence, neurosis and psychopathy which have hitherto been seen from a purely psychiatric angle. This modern version of the hypothesis that "the balance of neurotransmitters is a condition for normal behaviour" will surely give an impulse for further far-reaching research to be carried out.

Contents: Generic and Trade Names. – General Introduction and Definitions. – Sites of Action of Psychoactive and Treatment Strategies. – Pain. – Sleep. – Parkinson's Disease. – Depression. – Autonomic-affective Dysfunctions. – Neuroses. – Psychopathic Disorders. – Neurotransmitters in Old Age. – The Significance of Neurotransmitters for Human Behaviour. – Epilogue. – Bibliography. – Subject Index.

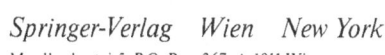

Springer-Verlag   Wien   New York

Moelkerbastei 5, P.O. Box 367, A-1011 Wien
Heidelberger Platz 3, D-1000 Berlin 33
175 Fifth Avenue, New York, NY 10010, USA
37-3, Hongo 3-chome, Bunkyo-ku, Tokyo 113, Japan

Supplementum 29

M. B. H. Youdim and K. F. Tipton (eds.)

Neurotransmitter Actions and Interactions

Proceedings of a Satellite Symposium of the 12ᵗʰ International Society for Neurochemistry Meeting, Algarve, Portugal, April 29–30, 1989

This volume, produced as a tribute to Professor T. L. Sourkes (Montreal, Canada) in the year of his 70th birthday, brings together contributions from established experts in neurotransmitter actions and interactions. Fields covered include the actions of peptide transmitters, amino acids and excitotoxicity and the functions and metabolism of the amine neurotransmitters.

*Springer-Verlag
Wien New York*

1990. 88 figures. X, 304 pages.
Soft cover DM 240,–, öS 1680,–
Reduced price for subscribers to "Journal of Neural Transmission":
Soft cover DM 216,–, öS 1512,–
ISBN 3-211-82142-2

Supplementum 28

G. D. Burrows and M. Da Prada (eds.)

Reversible MAO-A Inhibitors as Antidepressants

Basic Advances and Clinical Perspectives

This book offers state-of-the-art vistas on an exciting new class of antidepressant drugs, e.g. reversible (short-acting) and selective type-A monoamine oxidase (MAO-A) inhibitors. A well-assorted cohort of preclinical and clinical investigators illustrate several relevant aspects of the MAO-A, dwelling on their antidepressant effects in various subgroups of patients suffering from endogenous, exogenous or atypical depressive disorders. Because of their good tolerance and mild side effect profile, MAO-A inhibitors appear to have a wider gamut of therapeutic potential in psychiatry than tricyclic antidepressants.

The impressive array of evidence presented in this book convincingly demonstrates that the new generation of reversible MAO-A inhibitors are effective antidepressants with less side effects than available tricyclic antidepressant drugs.

*Springer-Verlag
Wien New York*

1989. 40 figures. VIII, 106 pages.
Soft cover DM 48,–, öS 340,–
Reduced price for subscribers to "Journal of Neural Transmission":
Soft cover DM 43,20, öS 306,–
ISBN 3-211-82133-3

Supplementum 27

J. A. Obeso, R. Horowski,
and C. D. Marsden (eds.)

Continuous Dopaminergic Stimulation in Parkinson's Disease

Proceedings of the Workshop, Alicante, September 22–24, 1986

The subject of this book is the outcome of a workshop held in Alicante/Spain in 1986 on the topic of "Continuous dopaminergic stimulation in Parkinson's disease" and consequently deals with a new continuous s.c. infusion method using the dopamine agonist lisuride for treatment of fluctuations in motor performance following long-term treatment with L-DOPA in Parkinson's disease which occur in a majority of cases. For galenical reasons, intravenous L-DOPA treatment can be given only for a short period, so it became necessary to devise new ways of achieving an easily controllable and constant dopaminergic stimulation, both for further research and treatment.

At this meeting, a number of clinical groups confirmed and extended the original findings by J. A. Obeso and his colleagues. These authors found that continuous s.c. infusion of lisuride, a watersoluble dopaminergic 8-α-amino-ergoline with dopaminergic properties which can be injected or infused, can improve—sometimes quite considerably—motor function in severely disabled fluctuating parkinsonian patients. The concurrent use of the peripheral dopamine antagonist domperidone attenuates or prevents side effects related to the stimulation of "peripheral" dopamine receptors, including the chemo-receptor trigger zone and some areas of the hypothalamus outside the blood-brain barrier. The clinical results discussed in this volume may not only be a basis for further improvements in our knowledge and therapeutic strategies in Parkinsonism, they point to the so far neglected importance of different ways of stimulating neurological or other systems, e.g. discontinuous, oscillatory effects caused by frequent oral application vs. continuous stimulation as described here with the lisuride s.c. infusion. Similar concepts have to be discussed and investigated in neurological disorders. In this respect, this multi-disciplinary meeting and its publication may offer new ideas and concepts for therapy in general, in addition to its potential application in the treatment of the complications of Parkinson's disease.

Springer-Verlag
Wien New York

1988. 58 figures. XII, 255 pages.
Soft cover DM 114,–, öS 800,–
Reduced price for subscribers to "Journal of Neural Transmission":
Soft cover DM 102,60, öS 720,–
ISBN 3-211-82034-5

Supplementum 26

M. B. H. Youdim, M. Da Prada,
and R. Amrein (eds.)

The Cheese Effect and New Reversible MAO-A Inhibitors

Proceedings of the Round Table of the International Conference
on New Directions in Affective Disorders, Jerusalem, April 5–9, 1987

This collection of papers gives the most up-to-date information
about the new class of monoamine oxidase inhibitor (MAOI) antidepressants which are reversible, preferentially inhibit monoamine
oxidase (MAO) type A and are non-hepatotoxic. One of these,
moclobemide (Ro 11-1163, Roche) is a benzamide derivative which,
in contrast to the classical irreversible MAOIs, is very unlikely to
cause the "cheese effect" (relevant increase in blood pressure, hyper-
tensive reaction) when tyramine (TYR) rich food is consumed
during moclobemide treatment. Thus patients need not adhere to a
TYR-restricted diet. Three papers show that moclobemide is an
effective antidepressant with a low risk of serious and minor side
effects. Four papers discuss in vitro or pharmacokinetic studies and
two papers discuss human studies, concerning the moclobemide /
TYR interaction. One paper describes the analysis of TYR content
of European food / drink since the existing literature is contra-
dictory. Summarizing, a psychopharmacologically-oriented psychia-
trist states that the risk of "cheese effect" occurrence was in the past
exaggerated even with the "classical" MAOIs and stresses the
clinical requirement for a drug like moclobemide.

Springer-Verlag
Wien NewYork

1988. 39 figures. VII, 136 pages.
Soft cover DM 70,–, öS 490,–
Reduced price for subscribers to "Journal of Neural Transmission":
Soft cover DM 63,–, öS 441,–
ISBN 3-211-82031-0